The Caravan and Motorhome Book

Collyn Rivers

RVBooks.com.au (2019)

The Caravan & Motorhome Book covers every conceivable aspect of camper trailer, caravan and fifth wheel caravan, campervan and motorhome usage. Like all of Collyn Rivers' books it is technically sound yet written in plain English.

Publishing Details

Publisher: RV Books, 2 Scotts Rd, Mitchells Island, NSW, 2430.
info@rvbooks.com.au

National Library of Australia - Cataloguing-in-Publication data

Rivers, Collyn

The Caravan & Motorhome Book

Third Edition

ISBN: 978-0-6483190-5-4. The Caravan & Motorhome Book. 1. Caravans. 2. Fifth Wheel Caravans, 3. Caravans & Motorhomes. 4. Motorhomes, 5. Camper-trailers. Title.

Editorial: Enquiries should be addressed to: collyn@rvbooks.com.au.

Front cover: Anthony Warry Photography, design by Collyn and Maarit Rivers.

Publisher's Note: To ensure topicality this book will be updated as required. The author would appreciate feedback relating to errors and omissions.

Chapter Listing

Acknowledgements

I thank the Campervan & Motorhome Club of Australia Ltd and CMCA staff throughout. I also acknowledge the UK Caravan Club and the National Motor Museum Trust (Beaulieu) for background information on early RV history.

I thank my wife Maarit Rivers for providing the section 'Keeping Well', also Glenn Portch for his input on lightweight construction (and friendship) and also those many individuals and companies who gave permission to reproduce their photographic and other material. Every attempt has been made to acknowledge copyright of photographs but we would be grateful for any errors or omissions and will correct them in the next edition of this book.

Dedication

This book is dedicated to the memory of my good friend Barry Powell (BSc.). Barry Powell was generous with his extensive knowledge and (since 2005) assisted in many ways in the writing of my books. He passed away (aged 73) in September 2015. He is still sadly missed by all fortunate enough to have known him.

Barry was a 'larger than life' man with strong opinions and ideals, and particularly a strong sense of justice and belief that all deserve a 'fair go'. Although not all agreed with him, almost everyone who knew Barry shared great respect and affection for him, not least as he appeared never to hold a grudge for anyone who did not share his views. Barry spent much of his working life in Canada teaching mathematics and physics, returning to Australia around 2002.

Widely known on RV forums as 'Budget Barry', he could do wonders with what most would see as scrap and converted a Nissan Civilian coach into a motorhome on that basis. Barry was also an accomplished bush poet and performer.

Preface

The *Caravan & Motorhome Book* originated in the early 1990s as an extended magazine series primarily related to campervans and motorhomes. It was expanded, updated and published in book form (as the *Campervan and Motorhome Book*) in late 2001 and updated between print runs. A (2011) second edition included fifth-wheel caravans, slide on units, and 4WD motorhomes as well as a great deal of all-new material. This was again fully updated in 2016 (to take in camper-trailers and conventional caravans), again in 2018 and once again for this edition (late 2019).

Among ongoing vehicle changes is the major acceptance of hybrid vehicles and the introduction of electric-only cars and 4WDs. It is also now all but certain that diesel-powered vehicles will be phased out worldwide (almost certainly by 2030), and possibly petrol-powered vehicles likewise.

Solar modules have continued to increase in efficiency and are still decreasing in price per watt.

Lithium (LiFePO4) batteries are still costly but well on their way, in many RV applications, to replacing conventional deep-cycle batteries. There are also many promising new battery technologies that will surface within the next few years.

There are increasing complications (with many new vehicles) in using alternator power for RV auxiliary purposes. It is likely future emissions regulation may preclude its use. Fuel cells, however, are an adequate substitute, as they are also for emission-producing generators.

Increasingly, apart for our books and website (https://rvbooks.com.au) next to no reliable unbiased information is available about buying, building, converting, or using camper-trailers, campervans, motorhomes and caravans. Most published product 'reviews' are now so-called advertorial, and included as part of paid advertising packages.

There are ongoing reports of varying and often misleading information, particularly about the declared Tare Mass (i.e. ex-factory weight) of caravans, and also of the size and weight tow vehicle required to tow them. Further, most caravans and motorhomes have grossly too little weight allowance for personal effects.

The *Caravan & Motorhome Book*, and its associated website https://rvbooks.com.au, ongoingly attempt to remedy this dearth of information. Both assist those seeking to buy, build, refit - and/or to plan the 'Big Trip'. As with our books (and websites), both are free of industry bias. None accepts paid advertising of any kind. All product pictures are reproduced at our expense.

This book should help you avoid the major traps prior to buying or building. Reader feedback is welcomed.

Collyn Rivers

Chapter 1

The early years

The concept of living in vehicles originated not from gypsies, but circus folk in France, around 1800. They used huge wagons that transported their circus props, and had basic living space for the performers.

Around 1810, Venetian ex bull-fighter and later showman, Antoine Franconi built, lived and travelled France in his 'voiture nomad'. It had two rooms and a large balcony. Charles Dickens' The Old Curiosity Shop (of 1840) has a fictitious Mrs Jarley who travelled the UK in 'a smart little house upon wheels.'

Romany people lived in tents and other forms of shelter until the late 1840s. Their first caravans (called 'vardos') were first seen around 1850. Vardo design soon became virtually standardised. They had large diameter wheels set outside the body that traditionally had the door at the front. It was not until 1900 or so, however, that most Romany folk (in the UK) had caravans.

The vardos' interiors had small cast-iron cooking stoves set into timber fireplaces, inbuilt seats, cupboards and wardrobes, plus bunks at their rear. Some had clerestory windows (ensuring light and air from above). Their finish varied from austere to superbly carved, often in brilliant colours or even sheathed gold leaf.

Vardos were often made by high-quality coach builders. They could take up to a year to construct.

In 1885, caravanning was boosted by the Salvation Army. It built caravans, each accommodating twelve young men plus an officer, to preach in villages and sell its *War Cry* magazine.

Recreational caravans

The first 'recreational' type caravan is believed to be the The Wanderer. It was built in 1885 for Dr William Gordon Stables. Aware that the Romany's major vardo builder was Bristol Wagon Works, Dr Stables ordered a 'gentleman's version' - of mahogany with a maple lining.

Figure 1.1. 'The Wanderer' - the first 'known gentleman gypsy' abode. It was owned by Dr William Gordon Stables, who travelled with his bagpipe-player/cook servant. Pic: Caravan Club UK.

The Wanderer was about 5.5 metres (18 ft) long and weighed about two tonne. It was pulled by two horses - and preceded by the doctor's valet on a tricycle.

His first trip, from Twyford (Buckinghamshire) to Inverness in Scotland, is described in his book *The Cruise of the Land Yacht Wanderer.*

Gentlemen gypsies

Some owners, by now called 'gentlemen gypsies' (despite about 10% of such caravans being owned by wealthy ladies), built even larger caravans. Some had hot and cold running water, libraries and even pianos. That, of the then Duke of Newcastle, had a large wine cellar.

Whilst mostly staying on properties owned by fellow gentry, their presence was not always welcomed on the roads. Dr Stables noted that stones were thrown at his caravan from time-to-time.

Horse-drawn reaction

Horse-drawn caravan usage resulted in the formation (in 1907) of the Caravan Club of Great Britain and Ireland. Caravan Club founder J. F. Stone, initially accepted owners of the then rare motor caravans and motor-towed caravans.

Figure 2.1. The 1926 Voyageur. Courtesy of Winchester Caravans.

He later felt that both were 'contradictory to the true spirit of the (horse-drawn caravans) pastime.' The resultant rift caused the Caravan Club to lose 95% of its members. It did not recover until 1935. Motor caravan owners were banned from 1914 until the 1970s.

A similar issue afflicted Australia's Campervan and Motorhome Club - that for over 25 years accepted conventional caravan owners only as non-voting Associate Members - yet fifth-wheel caravan owners as Full Members.

Motor-drawn caravans

The first known motor-drawn caravan was built in 1897. It was about nine metres (29.5 ft) long and two metres (6.6 ft) wide. Its railway carriage layout included a 22-seat salon/dining room, sleeping space for 11 people, a kitchen with an oil stove and icebox, and a toilet with hot and cold running water. It was pulled by a de-Dion Bouton steam tractor and said to have been in use until 1924.

Figure 3.1. Winchester caravan of 1936: arguably the finest of its era. Pic: Winchester Caravans (UK).

The 'gentleman gypsies' (10% were actually 'gentlewomen' often travelling alone) began to be socially challenged from 1921 when more affordable motor-drawn caravans from Eccles created a new market.

Eccles and more up-market caravans such as from Angela, Car Cruiser, Cheltenham, Hutchings and Rice nevertheless retained a shape and style influenced by the horse-drawn caravan era.

Figure 4.1. USA trailer park. Pic: BBC UK).

The 1930s saw a new approach. Almost suddenly, form switched to reflect function: compare the box-like Voyageur (Figure 2.1) made in 1930, and its successor from the same maker (Figure 3.1) built only six years later.

Prior to World War 2, only relatively well-off people could afford to buy a car, so caravanning by and large was mainly a pastime for the rich in what was then a very socially-stratified country. Many caravanning-related references of that era name titled people. Post-war, however, the UK caravan industry divided. One section continued to make recreational units, the other made mobile homes.

Britain's 1970 Caravan Sites Act, required the government to provide caravan sites. The Act was abolished in 1994. Local councils, instead, were then required to provide sites - but few (if any) did.

As in Australia, caravans then became used mostly for cheap holidays, often in hired caravans in caravan parks. A few owners used them for extensive touring, but were hampered by the-then few surfaced roads.

In the USA, a concentration on recreational caravans created caused social issues that exist to this day.

Vast numbers of so-called travel trailers began to be trucked to huge trailer parks that grew to become virtually small towns.

Figure 5.1. 1956 Airstream

The US Airstream company is believed to be the oldest still-existing caravan maker. It was started (by Wally Byam) in the late 1920s, originally selling kits for people to build their own caravans.

In 1936, Byam launched his 'Clipper' that was a look-alike Bowlus caravan. The product was superbly made and engineered and has retained its shape and alloy finish to this day (Figure 5.1). The company also builds fifth-wheelers and motorhomes.

In the 1930s, Australia's Paramount company made many classic caravans. That shown in Figure 6.1 is the four-berth (1937) unit. It was 15.5 ft long (4.75 metre) and weighed 15.3 cwt (under 800 kg).

Figure 6.1. The 1937 Australian-made Paramount. Pic: Original source unknown.

Later, in 1939, Australian manufacturer Carapark offered the all steel 'Superb'. The first was tested on a trip to Darwin, however WWII caused production to be mothballed.

The units began production post-war with the Carapark Hunter. These were jig-manufactured in batches of 18 (one batch every second day). Many were bought by British immigrants (for housing) on a special deal of 25% deposit and money back if not satisfied.

Figure 7.1. Often nick-named 'the Toaster' the Carapark was not the most elegant of caravans, but sold in substantial numbers. Pic: original source unknown.

Later, with the increasing number of small cars in Australia, Carapark released models such as the Hunter Minor (2 berth) and Hunter Junior (3 berth).

In post WW2 Australia, most caravan makers concentrated on mass production, with only a few making up-market products.

In the late 1990s UK/EU makers adopted lighter yet strong methods of design and construction. This began to be adopted by local campervan and motorhome makers (but less so by most Australian caravan makers). Some locally-caravans now weigh (unladen) 3000-3500 kg and a few over 4000 kg - yet may be towed by vehicles 1000 kg lighter. This issue is covered in depth in Chapter 7.

Fifth-wheel caravans

The first known fifth-wheel caravan is the 1917 (USA) Adams Motor Bungalow. (Figure 8.1).

Figure 8.1. The 1917 Adams Motor Bungalow. Pic: Glenn H. Curtiss Museum (USA).

It is believed to have been mostly designed by Adam's half-brother, the engineer Glenn Curtiss. By 1921, Adams was advertising five camping versions, and ten load-carrying commercial versions (i.e. early semi-trailers).

Little is known about the original unit but the later Curtiss fifth-wheeler caravans of the 1930s were spectacularly successful.

Figure 9.1. 1932 Curtiss Aerocar. The tow vehicle was a 1932 Graham-Paige 8-cylinder coupé. Pic: courtesy of the HET National Automobile Museum, Steueweg, 8 NL.

One Curtiss unit (Figure 9.1) doubled as a mobile office for financier Hugh McDonald's daily journey to his office in Manhattan, New York. The 'bridge' at the front includes a compass, altimeter, speedometer and barometer.

The trailer was occupied by its owner and his staff whilst being towed - it had a chef and his galley, a small dining area (and toilet).

The remarkable Highway Palace fifth-wheel caravan (Figure 10.11) was built in 1949 by the little-known Grace Bros (Adelaide) for Harry and Daisy Parr of Pannaroo (South Australia) but, for reasons that are unclear, was later scrapped. The caravan was later found at Murray Bridge The tow vehicle was found in a swamp in Wellington. Both were in a seriously bad state of repair but Dick Hart and his friends restored them over some five or so years.

Figure 10.1 The Highway Palace. Pic: pinterest - original source unknown.

The front upper part of the 13 metre (42 feet) enormously-heavy trailer is a dining area. There is a virtually home-sized kitchen and vast living area.

There is just one (but large) bedroom with two single beds, and an en-suite bathroom with a sunken bath, wash basin and toilet.

It is towed, via a 230 mm (9 inch) towball, by a 1940s Fargo. It is still driven occasionally but its sheer weight (and the inefficiency of engines of that time) results in fuel usage of a reported under 1.5 km per litre.

In the UK, a few fifth-wheel caravans were made by converting the ex-RAF Bedford 'Queen Mary' ultra-light semi-trailers used to carry aircraft during WW2.

Figure 11.1. This self-made fully off-road fifth-wheeler photographed in 1959 in the caravan park that, to this day, is in the Bois de Boulogne - in the centre of Paris. Pic: rvbooks.com.au.

Off-road fifth-wheel caravans

Fifth-wheelers are now rarely used off-road, but an interesting one historically is shown in Figure 11.1. The British owner (who had made it) towed it across Africa from Cape Town - including the 3000 km Sahara. It was was pulled by an Austin Gypsy 4WD - a vehicle notable for having five speeds - both forward and in reverse.

Motorhomes

The earliest known motorhome was built about 1905 by Belsize Motors for a J.W. Mallalieu (Liverpool). Of a then considerable 40 horse power, it weighed about three tonne.

In 1906 the UK Caravan Club member Albert Fletcher had a motor caravan built on a Rykneild chassis. It had a large seating area on its roof enabling those travelling to have a good view.

Figure 12.1. Albert Fletcher's 1913 Daimler-based motorhome. Pic: Caravan Club (UK).

Albert Fletcher later (1913) had a similar bodied motor caravan built on a 1913 Daimler chassis (Figure 12.1).

Another early pioneer was Madge Paton who, in 1912 had a motor caravan named the 'Tortoise'.

Figure 13.1. 1910 Pierce-Arrow Touring Landau. Pic: Original source unknown.

The first known production 'motor caravan' is believed to be the 1910 Pierce-Arrow Touring Landau. (Figure 13.1).

The chauffeur-driven vehicle had a backseat that folded to form a bed, a sink behind the driving seat and a chamber pot housed in a timber cabinet.

In the UK, commercial development was hindered by light commercial vehicles being limited to 20 mph (32 km/h) until the 1950s!

By 1957 only six UK companies offered motor caravan conversions. It was not until the late 1960s that commercial production seriously began.

Australia - motorhomes

Earlier Australian motorhomes may well have existed, but the National Motor Museum recognises engineer Gerhard (Pop) Kaelser's as the first.

Figure 14.1. Earliest known Australian motorhome is this unit built in 1929 by Gerhard Kaelser. Pic: Campervan & Motorhome Club (CMCA).

The house-like unit is on a 1924 Dodge truck chassis. It is about 4 metres (12 ft) long, by 1.9 metres (6 ft) wide, and has 3 metres (9 ft) headroom.

The interior has beds, table and storage. The driver and front passenger seat convert to a child's bed if required.

Gerhard Kaesler and his family used the vehicle to travel throughout New South Wales, Queensland, South Australia and Victoria.

The motorhome was later bought by the Mayor of Goolwa, and given to the District Council of Port Elliot and Goolwa after his death. It was later vandalised and then stored until the 1970s (when the council decided to dump it). It was later saved by a member of the local National Trust who arranged for it to stay in Goolwa Museum. In 2000, two volunteers restored its engine. It is now in working order and road registered.

Gerhard Kaelser later built a second one on a 1929 Dodge truck chassis. He recorded the construction of both in his book *The Beginning of Motor Caravanning in Australia*.

Australian campervan/motorhome production

Despite many campervan imports from the UK and Germany, large scale production did not really begin until the early 1970s. Then, LNC Industries bought Motor Caravan Holdings Pty Ltd and its subsidiary, Eric W. Sopru Pty Ltd, makers of the Sopru VW Campmobile conversion. Later Sopru campers were sold as 'factory VW products'.

Figure 15.1. The 1974 Westfalia Kombi in camp near Boulia (Qld). The author's wife (Maarit) is watching the local bird-life. Pic: rvbooks.com.au.

Whilst deservedly popular, the superb German-built Westfalia Kombi camper was never sold directly in Australia, but many were privately imported. One such (a 1974 unit with a pop-top roof) was owned by the author and his wife from 1994 to 2006 - and its sale regretted ever since (Figure 15.1).

The Dormobile campervan, first based on the Bedford CA, was built by a UK company (originally Martin Walter) formed in the 1700s! The company's Australian operation produced VW Kombi conversions offering VW warranty and parts and service support.

Figure 16.1 An early (1970s) VW Kombi conversion. Pic: Trakka.

Trakka, established in 1973, concentrated initially on VWs and Land Rover conversions.

The company has always been innovative. It introduced 12 volt compressor fridges, swing-out stove units, swivelling front seats and gas-strut assisted 'pop- top' roofs.

Trakka was probably the first to commercially offer solar (in 1985), and later diesel-powered hot water systems, and likewise ducted heating and cook-tops.

(Sadly, Dave Berry, Trakka's founder and its managing director died in mid 2019).

The Winnebago originated in the USA in the early 1960s. In the late 1970s the Australian company Knott Industries began using the Winnebago name on its Australian products.

The US Winnebago company took legal action to prevent further use of its name. The Australian Federal Court initially ordered the cancellation of Knott's ongoing usage of its name, but the ruling was overturned in March 2013. Knott Industries, however, had already established a new brand name (Avida) and ceased using the Winnebago name.

Winnebago, now under the supervision of Apollo Motorhome Holidays in Australia is listed on the Australian stock exchange and manufactures product in Brisbane.

Chapter 2

What can go where?

Excepting for much of Western Australia (WA) and the Northern Territory, (NT) most roads close to Australia's coast are bitumen. So is the 14,700 km of Highway 1 that extends all the way around.

There are two main bitumen roads from Adelaide and Melbourne to Brisbane. One is mainly coastal, the other inland. There is a good road from Brisbane to Weipa (about 300 km short of the tip of Cape York, north Qld) but it is a very rugged dirt track thereon.

Another good road runs from Brisbane to Mt Isa and across to Tennant Creek.

On the west coast there are two bitumen roads from Port Hedland to Perth. One follows the coast, the other (via Newman) is inland. It is 400 km shorter but not of any real interest from Newman onward.

There is a good bitumen road from Adelaide to Darwin, via Alice Springs.

In late 2018 the only bitumen east-west road is between Adelaide and Perth (about 2700 km) across the Nullarbor Plain in South Australia, (SA).

Whilst not shown on the above map, there is an east-west route from Boulia (Qld) via Alice Springs to WA. It is mostly dirt but is being progressively surfaced.

The north of WA and the NT apart, most of Australia's states have often interesting and lightly trafficked back-roads of which the surfaces are usually well maintained dirt. The east coast and Victoria in particular have many such roads. Some national parks have hard surfaced roads, but most (also state forests) have dirt roads of varying quality.

Figure 1.2. Australia's major roads. Pic: © Bogdanserbam/Dreamstime.com

Off-road

To RV vendors the term 'off-road' may be any road lacking a white line down its middle. In reality, 'off-road' varies from three-lane dirt highways, to where no track even exists.

Figure 2.2. The Pentecost River crossing (Gibb River Road) is most typically like this. When not, the track is normally closed to traffic anyway. Pic: Kimberly Kampa.

Vendors can mislead as many have zero related experience. As a result, front-wheel drive campervans barely able to traverse wet grass may be described by vendors as 'off-road'.

It is also an issue of familiarity. An outback bus run is routine for its driver, but an adventure for a first time owner.

The Kimberley's Gibb River Road is an icon for many four-wheel-drivers. Some proclaim 'we conquered it' but my wife travelled it routinely every month it was open to visit outback Aboriginal communities using a bog-standard Toyota Land Cruiser. The only mildly challenging part is the usually shallow Pentecost River crossing (Figure 2.2).

Figure 3.2. The Talavera Track (from Newman WA to Alice Springs) is a rarely-driven route. When we used it in 2009 we saw no other vehicle for three days. Pic: rvbooks.com.au.

Outside the wet season the more rugged standard caravans, towed by 4WDs, readily cope with Australia's Birdsville, Oodnadatta, and Strzelecki tracks. So too do well-made motorhomes and campervans with good ground clearance. If driven with care and in mechanically good condition, occasional such use is unlikely to damage them. Doing so regularly though almost certainly will. It may also invalidate warranty.

The main exceptions, that do require more rugged vehicles include the upper part of Cape York (north of Weipa), some tracks off the Gibb River Road, the upper north-west coast of the Kimberley and the Simpson Desert crossings.

All of the above have deep sandy stretches, severe corrugation and often sand dunes. Most are seriously corrugated. They necessitate a really sound 4WD and a truly outback-experienced driver.

East-west tracks in north and central WA have a few vehicles a day. The plus 1000 km Talavera Track (Figure 3.2), that takes in part of the Canning Stock Route, has so little traffic that when the author and his wife drove it in 2009 they saw no other vehicle for three days. It is, however rare not to meet any at all.

Seek current advice before using any of the above. Ideally travel with another 4WD and carry at least one week's supply of water.

Think hard about whether you truly wish to travel extensively on dirt roads, or just like the idea. We loved it, but the reality is dirt and discomfort.

Only a few of Australia's most interesting areas really require the full-on hard-core vehicle approach, and virtually all can be amply explored via professional tours in suitably rugged vehicles.

Vehicles required for serious off-road travel are covered in Chapter 10.

Corrugation

Corrugation (Figure 3.3) is virtually the natural state of dirt tracks travelled by wheeled vehicles. It is an ongoing series of traffic-formed ridges about 700-900 mm apart. The severity of corrugation depends on the nature of the surface, the amount and type of traffic and when the track was last graded.

Traversing very rough going requires extreme care but it is long-term driving over corrugation that eventually damages all but the most rugged vehicles and trailers. In particular doing so whilst carrying excessive weight typically causes structural and other damage.

Over such going it is essential to have top quality shock absorbers. These need renewing every 30,000 to 50,000 km and their rubber bushes regularly inspected and replaced as needed. It also assists by reducing tyre pressures (all round) by 25% or so - and keeping speed below 80 km/h (50 mph).

Figure 4.2. This track is typical of corrugation Australia-wide. Pic: original source unknown.

Chapter 3

Campervan, motorhome or trailer?

An RV may be sought for any number of reasons: from occasional holidays away from home to virtually full-time living and travelling. Most stay on hard-surfaced roads: it is mainly camper-trailer owners that travel Australia's dirt highways. Some travellers like the convenience of caravan parks but, and increasingly, others free camp whenever they can.

Choosing what to buy is best determined by travelling needs, but compromise is often necessary. Camper-trailers are liked by tent campers seeking more comfort and, as they are also less costly, by young people. For those with children, a camper-trailer's large annexe allows room to play.

If long-term travelling is in mind, be aware it may fall short of that hoped. Delay commitment until you've tried it for a time. Do not sell up (unless wealthy) as escalating prices may rule out re-buying.

Whatever the choice, compromise may be necessary: the vehicle may also be needed for commuting, for collecting kids from school, or carrying trade tools. Here, a campervan or dual-cab slide-on may suit.

Big motorhomes provide space and comfort but are cumbersome in cities. Fifth-wheel caravans provide more space at lower cost, but need a suitable tray-backed vehicle to tow them.

It may also be necessary to compromise on size. Parking a big rig at home is not always feasible and state laws preclude parking vehicles that exceed 7.5 metres in overall length (including a car with a trailer attached) in a built-up area for longer than one hour.

A common issue is that one partner may be less keen than the other about having an RV. Compromises tend to be agreed, but not all work. One partner may insist on a fully-equipped kitchen - but that can wreck an otherwise workable layout. Another may want heaps of loved books, a full set of body-building weights or other heavy and/or bulky bits and pieces that cannot realistically be accommodated within an RV's limited space and weight capacity.

Trailer benefits

A trailer's major plus is having separate living and driving components. If the tow vehicle breaks down, or needs servicing, the living component is still there to use. In the rare event that the trailer is immobilised, the tow vehicle is still available. Such flexibility is also handy for driving to local clubs for an evening meal (common in small country towns) and enabling a return after dark knowing the living accommodation is there on 'your' camp site.

Figure 1.3. The 7.15 metre (23.5 feet) Evernew 1000 is an iconic Australian caravan. It is far from light, but can carry up to 1000 kg. It needs a suitably heavy tow vehicle. (It is sold directly by its makers.) Pic: Evernew.

Caravans work well for long-term stays on sites because the whole rig need not be moved every time you go into town. Cost may also be a consideration. The total cost of a caravan and new tow vehicle is typically less than a motorhome or a fifth-wheel (or conventional) caravan's of equivalent living space. The tow vehicle is also more usable for everyday use.

Caravans typically have a long life: twenty years is not uncommon. Depreciation is low: the better ones sell for close to their initial buying price. Trailers tend to be used less than their tow vehicles, hence are subject to less wear and tear than an equivalent campervan or motorhome.

Trailer downsides

There are exceptions, but most caravans have surprisingly limited actual weight allowance for personal effects. That seemingly available is often far less than apparently specified.

Manually moving a caravan (particularly if twin-axled) or a camper-trailer is never easy, but electrically powered jockey wheels, or electric motors that drive individual wheels do assist.

Towing a trailer legally precludes towing anything else behind it. This will limit owners to a motorhome if they tow a large boat, mobile workshop, laundry, small car or a micro-light aircraft.

Whilst the tow vehicle is available for other purposes, for all but trailers under a laden 1500 kg or so, that vehicle is likely to be larger and heavier than required when not towing. It will also use more fuel.

Campervan/motorhomes

Campervans and motorhomes work well for long journeys with frequent stopovers because next to no setting up/packing up is required.

Whilst less true of small campervans, most motorhomes have a higher weight allowance than most mass-produced caravans.

This is invaluable if seeking self-sufficiency as that requires substantial solar and battery capacity, plus carrying a lot of water.

Single vehicle downsides

A single vehicle limitation is that if major work needs doing on that vehicle, alternative accommodation is required. That in remote areas is costly, scarce and usually uncomfortable. This is especially a problem if replacement parts are specific to your vehicle. They may take a week or more to obtain.

Figure 2.3. Large campervan - or small motorhome? Many such, as this Fiat-based Trakka Torino blur the lines between. Pic: Trakka.

A rarely considered downside is that if you leave the site temporarily 'your' site is likely to be taken by a new arrival. Some owners secure 'their' site by leaving a table or chairs, but it is not unknown for these to be stolen whilst they are away. A polite notice on a pole usually works.

A further consideration is that, unless four wheel drive (4WD), the longer and heavier motorhomes cannot access many secluded free-camping sites. There can also be space issues for big rigs in some caravan parks.

Security

Security is lower with caravans as it is not possible to drive off without going outside. Motorhomes and campervans are aided by it being harder to tell if they are occupied. They are also easier to drive away if one really does feel seriously threatened.

Unoccupied trailers are more obviously so, not least as the absence of a nearby tow vehicle likely indicates no occupant/s. A caravan is more easily stolen. Many owners install tracking units.('Safety' is covered in depth in Chapter 30).

Cooking -inside/outside?

One point to consider is that an RV usually changes one's lifestyle. This is particularly so with cooking. First time buyers, for example, usually see an oven as essential, but that oven is likely to be used for storage after a month or two.

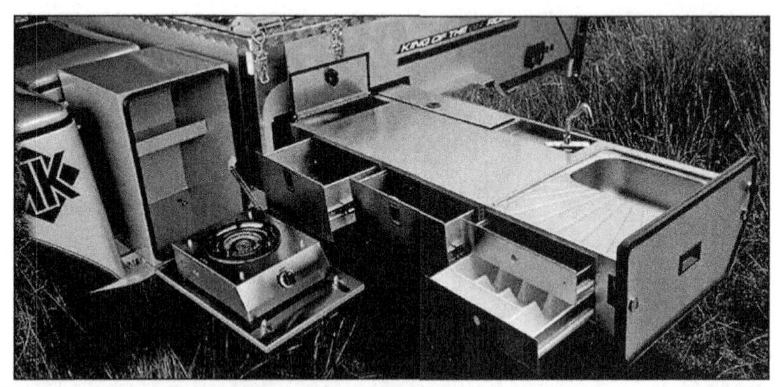

Figure 3.3. A second (slide-out) kitchen can be built into larger RVs. It slides in under a bed. Pic: Kimberley Kamper.

Most owners cook outside, and spend many evenings around campfires. This necessitates easy access to chairs and tables (one for cooking, another for dining) plus basic kitchen needs.

Likewise, washing-up is usually via a bucket - not the inside sink

Whilst an inside kitchen is handy when raining an all but essential inclusion is an additional (slide-out) camper-trailer style kitchen accessible from outside. Often needed items should ideally be equally accessible from inside and outside via duplicated doors or hatches.

The gas cylinder should be housed such that it does not have to be relocated and connected every time you set up the stove.

You'll need an awning that can be rolled out over the cooking area, plus provision for lighting this area. The awning needs to be quick to set up, not least because shade is surprisingly difficult to find in many areas away from the coast. It is virtually essential during lunchtime stops in the outback.

For most people, it is simpler, safer and more comfortable to sleep inside. This also simplifies setting-up and re-packing. Ensure you can access often-used cupboards and drawers with the bed in place.

Further considerations

Another, and major, consideration is where you intend to go. This substantially determines the type of vehicle you must choose. In practice, most serious off-road travellers use suitable camper-trailers or the more compact fully off-road 4WD campervans and motorhomes. Many 'off-road' caravans are made in Australia - but very few are seen off even the better-maintained major dirt tracks.

Also a determinant can be the type of driving licence required. If travelling with a partner, it is desirable that both may legally drive (thus both needing the appropriate class licence) - but not all may wish to do so. That and other legal matters are covered in Chapter 35.

Figure 4.3. The author's previously-owned OKA in Kakadu National Park - the spade marks where the bread is cooking. Maarit meanwhile checks the local bird-life. Pic: author 1994. Pic: rvbooks.com.au.

In practice, many RV owners choose and stay with whatever they first used and become accustomed to. They then see themselves as 'caravanners' - or 'motorhomers' ever after. Asking which is 'better' is like asking if quarter horses are inherently better than cart horses - it is RV horses for RV courses.

Chapter 4

Common requirements

Figure 1.4. Kitchen of the Avida Sapphire caravan. Pic: Avida.

As mentioned in the previous chapter, owning an RV is often only *one* partner's dream. The other may agree given a 'proper kitchen' or whatever, but this almost always results in unbalanced space and compromising practicability and resale value. In practice such kitchens are rarely fully used. An RV lifestyle tends to result in simpler and often healthier eating, and cooking mostly outside. It is very common to find that the originally seemingly essential inbuilt oven is rarely used.

RV builders catering for experienced buyers are aware of this. They provide that which they know most such buyers will find adequate. The kitchen shown in Figure 1.4 is typical (but, as noted above) an oven is unlikely to be used. Furthermore, layouts are usually *very* different from RVs intended for the rental market. Those typically have accommodation maximised at the expense of general livability.

Refrigerators

Camper-trailer and motorhome owners mostly use 12 volt electric compressor fridges, but many caravanners still prefer three-way units (230 volt, 12 volt dc, LP gas).

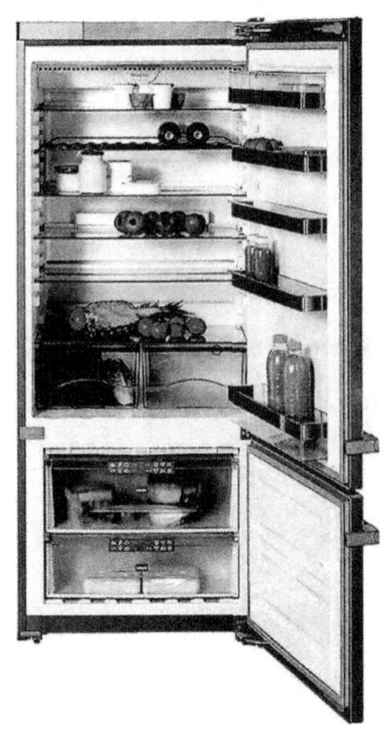

Figure 2.4. Miele fridge/freezer. Pic: Miele.

There is a growing move to ultra-efficient domestic fridges but (as with some RV fridges), some are wider than an RV's typical 600 mm entry door and accordingly installed prior to completion. Miele has a very efficient, albeit costly, 250 litre unit that is 580 mm wide (Figure 2.4).

Many fridges are obtainable as fridge-freezers, but a freezer is less necessary than a fridge as many food products are (or can be) vacuum packed. The associated machines are easy to use. Fridges are covered in Chapter 18. Their installation is covered in Chapter 11.

Beds

Some RVs have a bed/dinette/table arrangement that has to re-arranged into a bed each night. Not only does this become tedious but it does not suit couples who have (or develop) different sleep patterns, nor for anyone unwell. Many owners leave such beds made up during the day.

Figure 3.4. A cab-over bed is common with hire vehicles, but impractical for the less agile. Pic: Britz.

Many hire-fleet campervans and motorhomes have a bed over the driving cab accessed by a tall ladder. This is fine for the agile, but potentially dangerous for those not. Adding to the risk, the 'far end' sleeper must clamber over the other to access the ladder.

The elevated bed area adds weight in an area not originally intended to carry it. Weight there must be minimised.

This arrangement works very well, however, in fifth- wheel caravans. With these, the bed locates comfortably over the elevated hitch area, requiring only a few steps up from the main living area.

Some coach conversions have the bed on a raised platform that (in some) doubles as the top of a mini-garage. Another solution is a lightweight bed that rises and falls electrically. When not in use it is flush with a lower section of the roof.

Some beds fold into or against the rear wall, are in a slide-out section, or fold down from a partition that separates it from a rear toilet and shower.

Mattresses

Original RV mattress quality is not always good. Astute buyers negotiate a price reduction and supply their own.

Figure 4.4. Interior of a Paradise Integrity motorhome. Pic: Paradise.

The mattress base must be well ventilated or the mattress eventually becomes soaking wet. A pure wool blanket under the mattress helps keep it dry.

Toilets

This issue is to some extent age or agility related but, for RVs shorter than five metres (about 16.5 ft), consider whether it is worth sacrificing living and/or storage space to make room for a toilet and/or shower, particularly if planning to stay mainly in caravan parks.

An inside toilet is convenient but less necessary if free-camping. An outside toilet and/or shower is a workable compromise. Tents are made for this specific purpose. See also Chapter 16 re toilet chemicals. For a definitive technical opinion see our associated website's: rvbooks.com.au/napisan-a-professor-s-view/

Laundry

Whilst practical only in large RVs, a washing machine is worth considering. Those from RV suppliers are light and compact (Figure 5.4). A washing machine, however, far from essential. An hour or two's driving with dirty clothes in a plastic drum with a tight fitting lid, plus cold water washing powder is surprisingly effective.

Figure 5.4. This Lemair washing machine draws 300 watts and weighs 18 kg. Pic: Lemair.

Many travellers stay one night a week or so in a caravan park to do their laundry, shopping and possibly take in a local movie.

Heating

Most RVs now use diesel or gas space heaters (or combined water/space heaters). Most use a third or so litre of diesel fuel, (or LP gas) an hour. Also available are diesel or gas-fuelled storage water heaters. See Chapter 17.

Air-conditioning

There is a trend toward air-conditioning in the more costly motorhomes, campervans and caravans. A few even have non-openable windows. This is seemingly contrary to an RV lifestyle. Furthermore, moving into and out of air-conditioning in hot conditions severely stresses the body.

If you do use one, the reverse-cycle air-conditioners draw far less energy than others. Powering one all night nevertheless requires 400-500 amp hours of costly lithium batteries - weighing about 50 kg. Eventually, fuel cells will provide affordable power to cope with such loads. (Chapter 21).

Storage

Before planning starts, study the list in Chapter 27. You are unlikely to need all that included. For extended time away, allocate space and weight allowance for items bought during your travels.

For caravans, locate heavy items either over the axle/s or as close to it as possible. Never locate heavy items at the rear of any conventional caravan. Doing so seriously prejudices on-road stability - Chapter 7.

Floor coverings

Unless you drive only on bitumen roads and camp solely in caravan parks, avoid fitted carpet. It is a major effort to otherwise keep clean. Practicable alternatives include wood (bamboo is good and light) or high quality lino-leum, vinyl or cork, plus washable rugs. Wear Ugg boots if your feet are still cold.

Figure 6.4. Typical RV awning. Pic: Campervan Rentals Australia.

To minimise sand and dirt 'walked' into the vehicle, use doubled-up green shade cloth as a full length loose 'floor'. Dirt falls straight through the holes. This truly worthwhile tip was provided by a caravanning police ser-geant in Kakadu National Park in 1998.

Awnings

An awning not only shelters against sun and rain, it is warmer beneath an awning when cold. Some awning makers rate style over practicality. The more stylish an awning, the less it may withstand strong winds.

Full enclosure is bulky and heavy but zip-on or Velcro attached insect screens plus a 'skirt' to close the gap between the vehicle's floor and ground are essential in some areas.

Those travelling with children may use an annexe as their bedroom, but it limits its use for other purposes. An alternative for older children is a tent. Some can be erected and taken down in seconds.

Carrying firewood

Firewood is ultra-rare within 5-10 km of regularly used campsites.

Figure 7.4. Firewood carried on the rear of a TVan.
Pic: rvbooks.com.au

Collect it (early in the day) from areas recently cleared. Doing otherwise wrecks the habitats of creatures that live under long-fallen timber.

Be cautious when handling such timber: venomous creatures may live in it and resent being disturbed.

A volume of about 1000 mm by 200 mm by 200 mm is adequate for an overnight fire. That amount is shown here on the rear of our previously owned TVan (Figure 7.4).

Never load recently- burned timber: it may re-ignite whilst driving. For camper-trailers consider making a fold-down firewood rack of galvanised steel mesh.

Unrestrained objects

Secure everything that can be thrown forward in the event of an accident. Even at impacts at 20-30 km/h, unrestrained weight has an effective mass equivalent to many times its weight. A child died some years when a rigid foam freezer lid was hurled forward. It weighed under 2 kg, but the force with which it struck the child was as if it weighed 150 kg.

Travel lightly

Civilisation extends beyond freeways! You really do not need to carry several months' supply.

Most towns have a well-stocked supermarket and outback communities have general stores where you can buy most things - except fresh fish and even half-decent bread.

Even in the remote outback most Aboriginal communities have a well-stocked store (or mini-market) and will happily sell to travellers. Most can also supply fuel.

Chapter 5

Camper-trailers

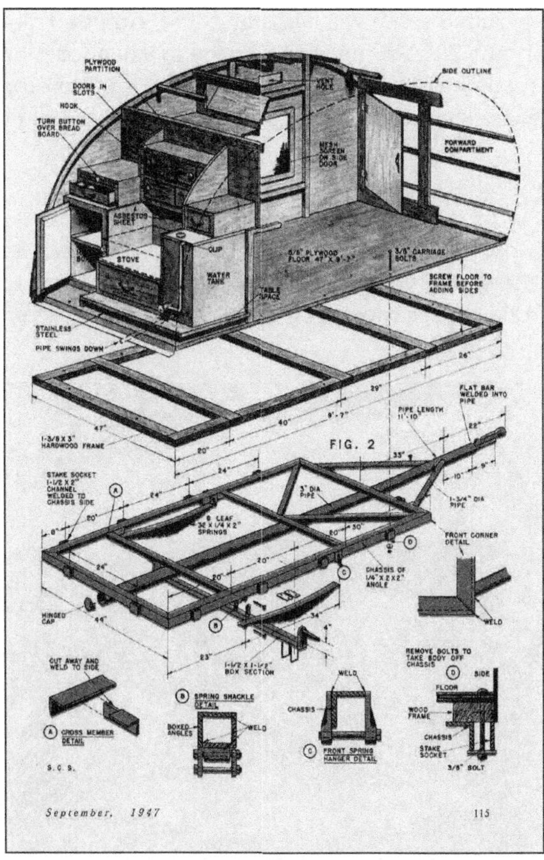

Figure 1.5. This is the 1947 Mechanix Illustrated 'Trailer for Two' unit made by Howard Warren of Riverside California. Pic. Photoshop enhanced by Caravan & Motorhome Books.

Camper-trailers began as self-build projects. Usage rocketed following detailed plans in a (1947) Mechanix Illustrated of a 350 kg 'Teardrop' concept (Figure 1.7). They have a following to this day.

Most camper-trailers are now larger and heavier: typically 1.9 metres wide, 3 - 3.5 metres long and of 500-1500 kg. The most basic is a box and tent on an existing trailer - with even a cheap kitchen being 'optional,' as are new (rather than part worn) tyres.

Most camper-trailers now sold in Australia are made in China. Many experienced users prefer locally made. Prices relate to quality and usage - not size.

Be very wary of imports in kit form. There may be good products out there but some are shockers. Don't buy without prior inspection.

For on-road units, the best value for money is $10,000 to $20,000. If shock absorbers are included some will withstand off-road use. Fully off-road camper trailers cost $25,000 upward. Optimum value for money is $25,000 to $35,000, but some cost almost $100,000! There is little that wears out, so used camper-trailers may fetch 85%-90% of their original price.

Soft floor

Soft floor units have a waterproof on-ground material that keeps out mud, and things that slide and crawl. Most have largish covered space, but may flood in heavy rain. They require flat, level ground that is often surprisingly hard to find.

Figure 2.5. Closable space offered by soft floor camper-trailers. Pic: jabiru.com.au

These units require time and effort to erect and pull down. Vendors insist it 'becomes quicker over time' but it may still take half an hour or more.

Repacking often necessitates firstly drying the canvas annexe (that usually stores over the bed). Water protection may be provided but wet canvas causes the trailer to stay humid, and smell for days.

Soft floor units are simple, light and usually cheaper. For some, that, plus the simplicity and large under cover area, may outweigh the downsides.

Hard floor

Hard floor units have a fold down section that forms a firm raised annexe floor. These trailers are usually quick to erect but are often heavier and usually more costly. They are less suitable for people with children as they have less covered space.

Figure 3.5. Hard floor camper: Pic: rangercampers.com.au

The Ultimate (Chapter 5) has a hinged platform section that forms a bed extension of the main body area. This enables dinette seating plus a kitchen and storage.

Kitchens

Most camper-trailer kitchens slide out from the trailer body, or from a drawbar-located enclosure. Soft floor trailers may have them free-standing or hinging out from the rear. Facilities vary from 'stove optional' and a few plywood shelves, to marine quality stove tops, grilles and stainless steel sinks.

Figure 4.5 Well above ground - the spacious interior and kitchen of the Ultimate camper. Pic: Ultimate.

Outdoor kitchens are great when the weather is fine but not when it is raining or freezing.

Protection from the wind is essential - without it cooking may be impossible as food and water may not be sufficiently heatable.

Overhead cover is also essential. Natural shade is rare across much of inland Australia. Overhead light is essential too. Yellow globes discourage most insects, but not all.

Have ready access to the kitchen, or carry a portable single burner cooktop for lunchtime stops.

Water

Most camper-trailers hold 60-120 litres. A viable alternative is the 'pure' water sold at fuel stops and supermarkets everywhere. This works well as safe water is scarce in outback areas and across much of South Australia and Western Australia.

Fridges

The fridge usually determines a camper-trailer's daily electrical draw: a 32 litre unit may draw 70% of that, an 80 litre unit may draw 90%. This topic is covered in depth in Chapter 18.

Interior heaters

The only truly safe forms of heating for use in an RV or annexe are the Webasto/Dometic type diesel-powered heaters, or Truma's LP gas equivalent (Chapter 17).

Energy overview

What works well for camper-trailers is to have two independent systems: one for the tow vehicle and another for the trailer. Two 100 or 120 watt solar modules on the tow vehicle's roof and a 100 Ah AGM, or lithium battery, will drive an 80 litre chest fridge in that vehicle with ease. A second system, of 60 to 80 watts, will run LEDs, water pump and a diesel water/space heater in the trailer.

Off-road camper trailers

There is an increasing trend to military-like design for fully off-road units that may weigh 1500-2000 kg. Many experienced users buy such units because they tend to be well-made and equipped. Their downside is that many are so heavy they require a heavy and powerful tow vehicle.

Figure 5.5. Fully off-road camper-trailer - the Conqueror UEV 490 has a Tare Mass of 1500 kg and ATM of 1950 kg. Pic: Conqueror.

These trailers do not necessarily need to be that heavy. The go-anywhere Ultimate is only 780-900 kg yet ultra strong. It is suitable for both on and off-road use.

Figure 6.5. A TVan camper-trailer can be set up for camping in less than 30 seconds. One part of the rear drops down to form a rigid floor, a second part forms a roof. If required an annexe drops down from the rear roof to enclose the elevated metal floor. A kitchen slides out from the other side. Pic: author's own rig a 2005 4.2 litre Nissan Patrol and 2005 Track Trailer TVan. Pic: rvbooks.com.au.

Track Trailer's 780 kg TVan, based on a chassis designed for military use, is small but works well once used to it's compact size.

For in-depth coverage of all aspects of camper-trailers please see *The Camper Trailer Book* (details in Chapter 38).

Chapter 6

Slide-ons

Figure 1.6. Early days -1968 Dodge Camper. Pic: J P Joans & allpar.com

The slide-on camper is a logical and practical way of enabling one vehicle to be truly dual-purpose. That vehicle may be a conventional tray-back for work etc, and a camper when required. Made famous in John Steinbeck's (1960) book *Travels With Charley,* it works best with the Ford 150 type US-made vehicles, but slide-on campers are made locally for Land Rovers and Toyota 4WDs etc.

There are two main types: those similar to small caravans (often with a pop-top roof), and those that are more compact but open up much as do camper-trailers.

Figure 2.6. This 1970s slide-on is an early Knott Industries Freeway product. Pic: source unknown.

There are also slide-ons that are as compact (but heavier) as hard-top canopy versions of the base vehicle.

As with smaller camper-trailers, slide-on units appeal to people who fish, and those seeking to explore remote areas.

Pic: Figure 3.6. Ozcape.com

A major issue with some slide-ons is excess weight. Many weigh 750 kg (or more) when unladen. One that RV Books became aware of, fitted to a Toyota Hi-Lux for which it was claimed to be suitable, was put over a weighbridge by the intending buyer. The unit (unladen) proved to be *already* over its legally permitted on-road weight.

Most slide-ons have retractable or folding legs that support them above ground level when not on the vehicle (Figure 3.6).

Most current units use strong lightweight materials - but not all do. Intending buyers are also advised to check that adequate allowance has been made for water, gas, and typical personal effects.

Compact slide-ons can be good value for money. They are an acceptable solution for those seeking a compact rig.

Figure 3.6. The Metalink SLR slide-on. Pic: Metalink.

Chapter 7

Caravans

One of a caravan's major benefits is flexibility. It can be left in a caravan park, or a town's outskirts whilst its owners go shopping, or exploring out-back tracks in their four-wheel-drive tow vehicle. Unlike a motorhome, it can be lived in whilst the tow vehicle is being serviced or repaired. Probably because of this (and their generally lower price) of all RVs in Australia (612,767 in 2018) about 90% are caravans and camper trailers.

Figure 1.7. The huge tri-axle Kedron.

Most are locally made, with Jayco having about 53% of the market. An increasing number of UK, EU and Chinese makers also and/or, assemble caravans in Australia.

Figure 2.7. A typical high-quality Evernew caravan - at an affordable price. Pic: Evernew.

Most caravans cost $35,000 to $100,000. Custom-made units cost more: the huge three-axled 8.4 metre Kedron (Figure 1.7) weighs 3990kg or 4490kg (depending on payload - of 620 kg or 1120 kg) and costs about $250,000, but companies such as the long-established Evernew build excellent custom-made products for far less.

Size and weight

Until the 1960s most caravans had a body length of 4-5 metres (15-17 ft) and weighed about 170 kg/metre unladen.

The average EU caravan unladen weight is currently about 210 kg/metre. The 5.8 metre (19 ft) Adria is 1171 kg unladen, and 1500 kg maximum on-road weight. Bailey's 2014, 5.68 metre (18.8 ft) Pegasus Verona - Figure 3.7. weighs only 1360 kg unladen.

Most Australian-made caravans weigh a lot more. The average of twenty on-road products (in 2019) was an unladen 350 kg/metre. Australian-made off-road caravans on average weigh 455 kg/metre (which largely explains why they are only rarely seen off-road).

Many local caravanners and caravan vendors still maintain caravans must be heavy 'to cope with Australian conditions'. This claim cannot be taken remotely seriously. Much of such weight is of non-structural medium density fibre board (MDF) etc used for the floor and often interior fittings. An MDF floor of a 7 metre van (21 ft) may exceed 385 kg, necessitating a correspondingly heavier chassis and suspension to support it. Lighter local products and EU imports have long proven their capability. There is a welcome move to lighter composite materials, but far from all makers use them.

Overweight issues

The Tare Mass quoted by a caravan maker (as it leaves the factory) excludes water (typically 180 kg) and gas (8.5 kg per 9 kg bottle). Many caravan's true Tare Mass is often much higher than declared. This is because, even if specified when placing the order, most caravan makers produce a basic 'standard' product and leave it to dealers to supply and fit all 'optional extras'. Not all dealers advise that the weight of items such as a second gas bottle, extra water tanks, batteries, solar, air conditioning, TV, washing machines, etc. is unlikely to have been included as Tare Mass - often leaving next to no allowance for personal possessions etc.

Figure 3.7. Bailey's 2014, 5.68 metre (18.8 ft) Pegasus Verona, weighs 1360 kg. Pic: Bailey.

Before signing the cheque, have the unladen caravan weighed (in your presence) on a certified weighbridge. Insist on being there for this - to prevent mattresses, drawers etc. being removed whilst weighing.

The maker must also advise the Aggregate Trailer Mass (ATM): i.e. the maximum the fully laden caravan may legally weigh when standing on a public road and not coupled to the tow vehicle. The difference between the (true) Tare Mass and the ATM, i.e. the 'Personal Allowance', is everything else fitted to or loaded into that caravan.

A long standing industry recommendation (not legal requirement) is that such allowance be 250 kg for single axle 'vans and 350-400 kg for dual axle caravans. More costly (and off-road) caravans usually have more, but much of that is the content of larger water tanks, additional batteries, etc.

For almost all mass-produced caravan, even if the Tare Mass is correct, the personal allowance is inadequate. e.g. the caravan's water tanks are included - but not the (1 kg/litre water content). Nor the 8.5 kg LP gas.

When buying, obtain a signed agreement that specifies the personal effects allowance that will be truly available for use as delivered.

If planning to free camp, spell out what is required in writing - as there can be issues with inadequate solar capacity etc. Few mass-produced caravans cope with more than one night away from 230 volt power.

See also Chapter 21, and our specialised books *Solar That Really Works!* and/or *Caravan & Motorhome Electrics* re this.

The most common complaints allege dealer refusal to accept warranty liability, and/or that owners may be told to contact the maker. The customer can legally choose to approach either, but ultimately it is the dealer, not the manufacturer, who bears final and legal responsibility for warranty issues.

Caravan stability: effects of weight

Any caravan towed via a hitch located behind the tow vehicle's rear axle imposes weight, yaw (snaking) lifting, pitching and rolling forces on itself and on that tow vehicle, requiring that vehicle to be (relatively) heavy enough to resist those forces. Not all are.

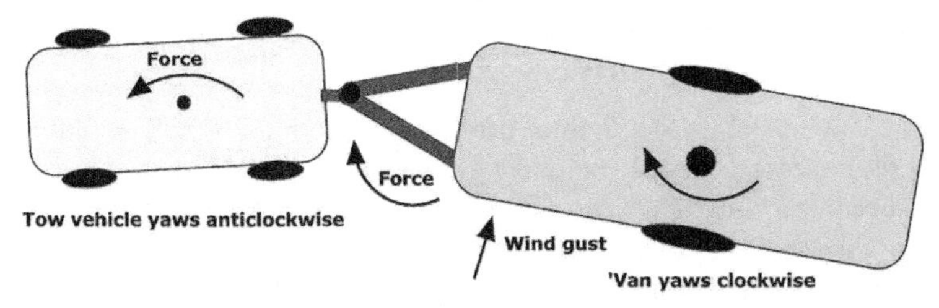

Figure 4.7. A yawing trailer towed via an overhung hitch causes its tow vehicle to yaw in the opposite direction, and vice versa. The greater the hitch overhang the greater the effect. Pic: rvbooks.com.au.

Australia's recommendations about tow vehicle weight often stem from the 1930s when few rigs exceeded 80 km/h. It was generally agreed (and followed) back then that trailers should weigh no more than the tow vehicle, but some authorities set a legal limit (but not recommended) of 50% more. Tow vehicles have since become increasingly powerful (yet lighter). Trailers are now longer and heavier.

In recent times organisations and authorities in many countries suggest (some legally decree) that laden caravans should not exceed 85% of the tow vehicle's unladen weight. One suggests 'those suitably experienced may go up to 100% of such weight, but that no-one should tow a caravan that is heavier than the towing limit of the vehicle it is behind'. Many countries have a towing speed limit of 80 km/h.

In 2013 the Caravan Council of Australia (CCA) backed by the-then Caravan & Motorhome Books (now RV Books) recommended that laden caravans should ideally weigh no more than 77% of the laden tow vehicle. Despite the CCA's recommendations being less stringent than in Europe, the recommendations resulted in user and dealer backlash, but many overlooked that the CCA referred to the *laden* weights of caravans and tow vehicles.

Australian legislation regulates that Gross Combination Mass (combined maximum weights) of the laden tow vehicle and laden trailer be set by the tow vehicle maker and must not be exceeded. The GCM is currently (2019) typically 5000-6000 kg (one Toyota model is higher).

Tow ball mass

For towing stability it is vital that caravans be front heavy. Based largely on what seemed to work (for short, light caravans in the 1930s), European makers suggested it be about 7% of the caravan's laden weight. Australian makers felt that 10% was more suitable.

By the 1960s '10% on the ball' had virtually become a mantra. Engineering laws, however, dictate it is not so much the *amount* of weight - but the *effect* of that weight that matters. And that is a function of the distance various weight is from the caravan's axle/s. A centre-heavy caravan is far more stable than an end-heavy caravan.

Industry tow ball mass 'recommendations' changed in 2012. Due almost totally to increased torque (i.e. pulling power), many 4WD makers increased claimed towing weight, yet reduced maximum allowable tow ball loading. Towing weight, however, relates to drawbar pull and/or ability to restart on a hill etc. Some caravan owners appear to not realise that caravan towing via an overhung hitch is a tiny percentage of the tow vehicle market. The weight a vehicle can actually physically support (or control) when towing is usually less - and dropping.

Figure 5.7. How trailer rear-end weight can affect a trailer. Had that load been centralised just ahead of the axle, the unfortunate animal would still have control. Pic: Source unknown.

Despite this, a few years ago many vehicle makers reduced chassis thickness from 3.5 mm to 3.00 mm, and their maximum tow ball mass accordingly. Many caravan makers reduced their 'advised tow ball mass' virtually overnight - yet with no engineering changes.

Now, many almost identical trailers have recommended tow ball mass from 4.2% to 10%.

Legal liability limits this book to advising to do whatever the vendor recommends. It does not, however, endorse such advice because most of the considerable research in this field advises that tow ball mass be 10%.

Weight distributing hitches

The tow ball load on an overhung hitch levers up the front of the tow vehicle thereby reducing loading (and grip) of the front tyres. Local and US practice has long been to use a weight distributing hitch (WDH) to lever the tow vehicle's front wheels back down (Figure 6.7).

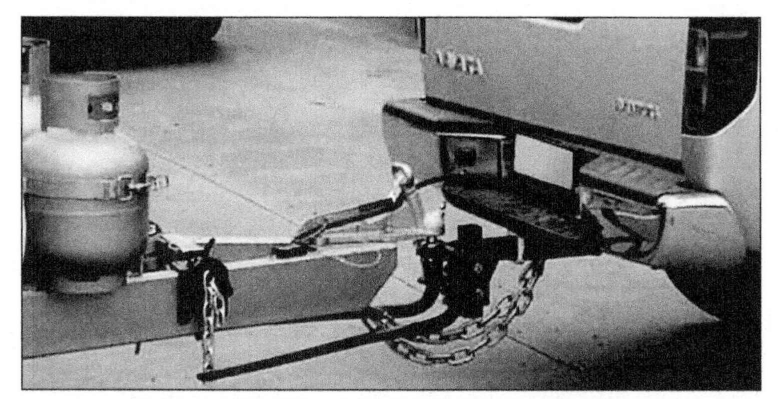

Figure 6.7. Ezy-Lift weight distributing hitch. Pic: Jayco.

Whilst a WDH controls weight effectively, its downside is that it cannot and does not reduce the yaw forces on the tow vehicle's rear tyre footprints. This introduces a rear-wheel steering effect that can prejudice towing behaviour. A full general and also technical explanation of this is in our associated book: *Why Caravans Rollover - and how to prevent it.'*

A WDH is in effect a truss attempting to compensate for maker/user self-inflicted excess end weight.

The EU's ongoing reduction of this weight is far more logical: few EU-made caravans have a WDH or even provision for fitting one. This approach minimises the total mass and particularly the end mass, and need for heavy tow ball mass. This increases the tendency to mild yaw but the forces involved are low and readily handled. Much local product is so nose-heavy, however, there is little choice but to use a WDH.

Most local advice is to adjust the WDH such that the tow vehicle's rear droop when the caravan is hitched up is fully corrected. Advice from a major (US) WDH maker, and now the US vehicle industry generally) is to correct to only 50% - typically resulting in the tow vehicle's front wheel arch lifting by 50 mm when the laden caravan in hitched up. Never attempt adjust until the caravan is level - as that needs 150%.

Sway (yaw) control

A caravan and its tow vehicle interact. Unless towed by an infinitely heavy vehicle, any caravan towed via an overhung hitch inherently tends to yaw. Each 'vehicle' causes the other to do so in the opposite direction (Figure 7.7).

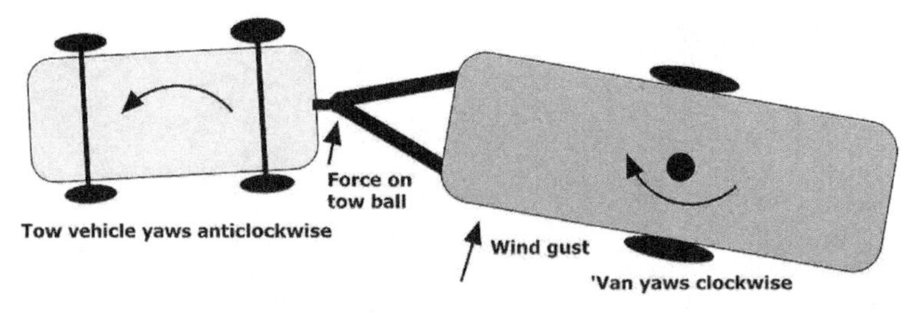

Figure 7.7. Caravan yaw. Pic: rvbooks.com.au

As long as such yaw dies out of its own accord (i.e. without driver-correction) within two/three cycles, such yaw is annoying but usually harmless.Sway control (the industry refers to yaw as 'sway') reduces it such that it is only barely noticeable. One type uses friction to dissipate yaw energy as heat. Another uses spring loaded cams that lock the caravan and tow vehicle together, relying on tyre distortion for cornering. The cams release only for very tight turns but may do so (not usefully) in an emergency swerve).

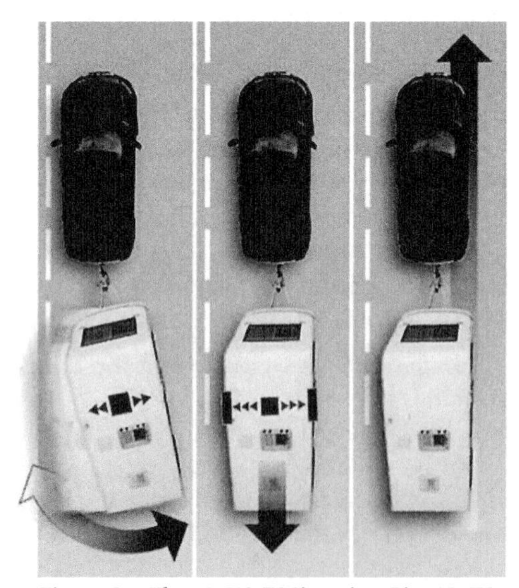

Figure 8.7. The AL-KO ESC's action. Pic: AL-KO.

Sway control should not be used to control yaw that, without it, does not self-limit after two to three cycles. Doing so masks instability that (at high speed) no friction or cam sway unit can control. Fix any such issues first and use that sway control only to limit discomfort at low speeds.

Electronic stability control

Many current tow vehicles have electronic stability control that assists to stabilise that vehicle alone. It may reduce, but does not necessarily remedy, trailer yaw.

Electronic stability control units made for caravans detect yawing and apply the caravan brakes only. The Tuson (also sold as Dexter) unit brakes the caravan's wheels individually and at lower yaw levels. This results in a calmer low speed drive, but (in the author's opinion) may mask instability issues better resolved at source.

Figure 9.7. Never extend hitch overhang. Pic: source unknown.

AL-KO ESC detects yaw and, if at a level deemed excessive it automatically applies the caravan's brakes for a second or two. It does so at about 75% of maximum braking, repeating the cycle if necessary (Figure 8.7). This system does *not* operate at low yaw levels. The units can be retrofitted to most caravans with AL-KO brakes.

Each system has its benefits, but as a generality, experienced caravanners tend to prefer the AL-KO 'parachute' approach as they are likely to have already resolved lower speed yaw issues.

Tow vehicles

Caravans need a tow vehicle of at least equal weight (ideally more), and with a short distance from tow ball to the centre line of the tow vehicle's rear axle. The 2019 average of such distance is 1.24 metres, but some Land Rovers have far less.

Figure 10.7. "I said that ute will pull it – I didn't say it should!" Pic: agcoauto.com

Be wary of utes (particularly dual cab units) with extended trays - that extension substantially extends hitch overhang.

Off-road caravans

Many caravans sold as 'off-road' are really intended for 'occasional use on good dirt roads'. Their suspension is usually a good indication. It matters not whether it is independent or beam-axle but it must be well made and have high quality shock absorbers. Ground clearance needs to be plus 250 mm.

Figure 11.7. One of the few truly serious off-road caravans - the Barry Davidson 1990s Phoenix. Pic: Phoenix.

For extensive off-road use, tyres need to be sized such that their on-road air pressure is not higher than 300 kPa (about 40 psi). This enables pressures to be reduced to the desired 140-210 kPa (20-25 psi) for crossing soft sand.

Some owners tow 6 metre caravans across inland tracks but 4.5-5 or so metres (about 14-16 ft) is more realistic. All need heavy and powerful tow vehicles. It is strongly recommended to have a front-mounted 7000 kg winch.

Buying second-hand

Caravans have a long life-span: their average age Australia-wide is over 17 years. They retain their value as there is little apart from shock absorbers, brake linings and and wheel bearings that actually wears out. Some people buy old ones and refurbish the interior.

Electric tow vehicles

It is now (late 2019) certain that, to reduce emissions, petrol and (particularly) diesel-engined passenger cars and 4WDs will be increasingly phased out - with none made after 2030. Many makers already have hybrids that run on electricity for 30-50 km or more before switching to fossil fuel.

Right now (late 2019), because coal and oil-fuelled electric power stations currently produce as much pollution as does a petrol or diesel-powered vehicle, an electric powered tow vehicle only reduces air pollution if their main energy source is wind or solar.

That too is changing as many homes with grid-connect 6.5 kW solar power have far more input than they (and the major electricity suppliers) need. Many such home owners are already using the surplus to charge small electric cars (despite such cars' currently [2019]) very high price.

The motor industry worldwide is, however, planning to launch affordable all-electric vehicles as from 2022. This includes 4WDs suitable for towing (a few makers already have them).

Contrary to much forum speculation, that an electric tow vehicle will lack torque (i.e. low speed power) the very opposite is true. Many electric motors produce their maximum torque at zero speed.

Providing the desired range will require substantial and heavy battery capacity (possibly backed-up by fuel cells - Chapter 21). For towing, this is a plus - as stressed in this book, the tow vehicle really does need to be heavier than whatever it pulls. Hybrid and all-electric tow vehicles have an edge here - as their batteries inevitably add a great deal of weight.

Caravan summary

Stability issues mainly arise when towing long end-heavy caravans by too-light tow vehicles and/or at high speed. The risk is lower with caravans under 6 metres towed by a heavy long-wheelbase 4WD with short hitch overhang, and correct loading. Problems manifest primarily when having to swerve very hard at speed etc. Owners of suitably matched rigs driven in a sane manner are unlikely to experience instability issues as long as they do not exceed 100 km/h.

A move to the inevitably heavier electric tow vehicles will assist - but is a long-term solution as the average of motor vehicles (according to the Motor Vehicle Census, Australia, 31 Jan 2019) the average age of our 19.5 million registered vehicles across mainland Australia is 10.2 years (and in Tasmania 12.9 years).

Totally overcoming caravan instability necessitates zero hitch overhang. If seeking a caravan longer than 6.5 metres (about 20.5 ft) and likely to weigh over 2.5 tonne laden it is worth considering one of less weight, or towing by a heavy vehicle (e.g. a Ford 150 or Dodge Ram). Or consider the fundamentally more stable fifth-wheeler alternative - Chapter 8.

Chapter 8

Fifth-wheel caravans

A fifth-wheel caravan is attached to its tow vehicle via a hitch that, if directly above that vehicle's rear axle, results in it being largely immune to side forces such as wind gusts etc. Weight distribution too is far less critical. Handling is similar to a motorhome's except that a fifth-wheel caravan adopts a tighter radius whilst turning. Turning circles are smaller: some rigs can turn with the trailer at a right angle to the tow vehicle. Fifth-wheel caravans are also much easier to reverse than conventional caravans. Because the space above the drawbar is utilised, a fifth-wheeler has about two metres more usable space than a conventional caravan and tow vehicle of similar on-road length. Care is needed, however, on winding tracks, particularly those with rock faces or steep drops at the side, as a fifth-wheel caravan does not turns in the same radius as the tow vehicle.

Tow vehicle

All fifth-wheel caravans need a tow vehicle's tray long enough for the trailer's nose to be clear of the driving cab when turning. The tow vehicle must, within its Gross Vehicle Mass (Chapter 36) be able to accommodate the nose weight of the trailer mass as well as that of fuel, driver, passenger/s, etc. That nose weight is typically 15%-25%, extensive research, however, has established that 10% or so has little if any effect on stability.

Figure 1.8. A well positioned hitch. Pic: source unknown.

Smaller fifth-wheelers can be towed by Holden and Ford utilities, extended chassis Nissan Patrols, or Toyota LandCruiser based tray-backs. Larger ones need a small truck - (e.g. the Iveco). Four-wheel-drive is preferable as the extra traction and low range gearing is valuable when hauling several tonnes up steep wet hills.

The rig's overall weight must not exceed the tow vehicle's Gross Combined Mass (Chapter 36). If that is 5.5 tonne and the tow vehicle is 3 tonne, the laden trailer must not exceed 2.5 tonne.

Hitch positioning

For optimum stability, a fifth-wheeler's tow hitch needs locating above the rear axle/s (Figure 1.10), but there must be space ahead of the hitch to allow clearance whilst turning tightly (eased by having a rounded front section). There are hitches that slide diagonally rearwards for turning at low speed, but as this moves weight to the side, such hitches need using with care.

Figure 2.8. The tow hitch must allow rocking at least as much as does this double oscillatory unit. Pic Frans Hamer.

Double oscillatory hitches

A fifth-wheeler twists, rocks, and turns relative to the tow vehicle. Unless the trailer's chassis and body enables it to flex (as with commercial semi-trailers) this rocking necessitates a tow hitch known as 'double oscillatory'.

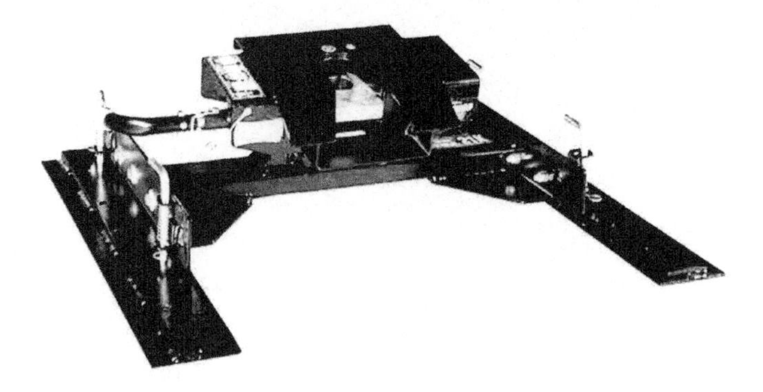

Figure 3.8. Single oscillatory hitch allows rotational and fore and aft movement but precludes sideways rocking. Pic: original source unknown.

Prior to 2010 or so, a number of fifth-wheelers, mainly imports but also some locally made, were inexplicably supplied with 'single oscillatory' hitches (Figure 2.8) that prevented the trailer rocking sideways.

This resulted in over 300 reported cases of cracked hitches and frontal section of trailers. In some cases it also damaged the trailer bed mounting. In 2014, it was found that a number of such hitches were *again* being used.

Figure 4.8, of the Australian-made D'Angelo double oscillatory hitch, shows the fundamental difference - of additionally allowing side-to-side rocking.

If considering buying a used fifth-wheeler, have a certified engineer check for structural damage. None may be visible, but metal fatigue has likely to have set in, but not visually obvious.

Suspension

Fifth-wheeler suspension should be similar to that of the better-made conventional caravans. They are fine for on-road travelling, but some makers omit shock absorbers, claiming they are unnecessary with the leaf springing mostly used. The omission of shock absorbers results in a harsh ride that may damage the trailer and its contents, particularly on rough roads. A few fifth-wheelers have AL-KO rubber suspension - that is inherently self-damping.

Figure 4.8 The locally made D'Angelo double oscillatory hitch. Pic: D'Angelo.

Figure 5.8. Australian-made Winjana fifth-wheeler off-road caravan. Pic: Winjana.

Winjana in Queensland also make off-road versions - Figure 5.8.

Imported fifth-wheelers

Privately importing a fifth-wheeler from America or Canada may initially appeal as it can be far cheaper than those brought in by established importers but, as some buyers find later, like is not necessarily being compared with like.

Figure 6.8. Typical high quality US fifth-wheeler - the Montana. Pic: Keystone (USA).

Whilst excellent value for their (US) price, many of these imported fifth-wheel caravans are very heavy (that shown in Figure 6.8 is over 7.0 tonne). To reduce nose weight, their axles are located close to their centre, resulting in an uncomfortable 'shunting' action whilst driving unless towed by a suitably heavy vehicle. (as they usually are in the USA). The discomfort can be overcome by using an air-sprung hitch - but it is better to have a lighter unit with rear-located axles.

There *are* many well-made US and Canadian fifth-wheelers. If bought with caution, following impartial advice, mid-price range imports are mostly fine, but any low-end fifth-wheeler import should be treated with considerable caution. Be ultra-careful about buying any privately imported RV, new or second-hand, (they are mostly fifth-wheelers) as there can be major problems with compliance. Few claimed to be 100% compliant actually are.

Most locally-built fifth-wheelers are lighter, enabling a desirable 20%-25% of their weight to be carried by the towing vehicle, and their axles to be close to the rear to aid stability.

By far the best was a limited run of ultra-light and superbly-made (ten only) units made on a not-for-profit basis by the now retired Glenn Portch. He retained one (Figure 8.8) for himself.

Fifth-wheel caravans still have a loyal following in Australia, and whilst sales were low in 2018 they are now (late 2019) claimed to be increasing.

Figure 8.8. This 'state of the art' 11.3 metre fifth-wheeler weighs only 3200 kg and has a 1300 kg payload. It is pulled with total ease by its 3 litre Iveco tug. A later 9.3 metre unit weighed 2350 kg (it could have been built weighing under 2000 kg but the buyer insisted in marble bench top inclusions). Pic: by designer/builder Glenn Portch.

Chapter 9

Campervans & motorhomes

Until 2005 or so most campervans were converted delivery-vans and motorhomes were purpose-built, but the distinction has now become fuzzy. Some campervan builders still retain the body of an original van, but add opening windows and often a pop-top roof. Others use the cab, engine and transmission of a front-wheel-drive delivery van grafted onto a light AL-KO chassis. These are imported as back-to-back pairs (Figure 1.9).They are then separated and and have a basic AL-KO chassis added by the RV maker.

Figure 1.9. Ducato front ends are imported as bolted-together pairs. Each has the front-wheel-drive engine, and gearbox) Pic. Fiat.

Such vehicles are typified by companies such as A'Van, Avida, Jayco (Figure 2.9), Sunliner and Trakka.

Some of the products have slide-out sections offering opened out on-site space without large on-road size. They are easy to drive and park, but their small tyre size and low ground clearance limit their usage to bitumen roads.

Motorhomes

The larger campervans/smaller motorhomes are based on light truck chassis: e.g. Isuzu and Mazda. Most are well made and good value for money. Their downside is that their ride may be harder than expected, but this can be partially overcome by converting to air suspension. New ones cost from $75,000 up, but good second-hand units can be bought from $30,000 upward.

Figure 2.9. Fiat-based Jayco Conquest uses an AL-KO chassis. Pic: Jayco.

Conversions based on the heavier Toyota Coasters and Nissan Civilians are good compromises between affordability, mobility and comfort. Both are ultra-reliable. Expertise, spares and service are available worldwide.

Old ones fetch $15,000 upwards but good ones 10-15 years old are $25,000 to $35,000. About $50,000 to $75,000 should buy one more recently made.

Figure 3.9. This is a good example of a professional (long wheelbase) Coaster conversion. Pic: Australian Motorhomes.

Toyota Coasters are available in standard and long wheelbase versions. Caution is needed when buying, as like many such vehicles, some have gearing intended for town driving. These have good acceleration from rest, but the engines rev their hearts out at touring speeds. This can be remedied by changing the differential (the main rear axle component) but this is likely to cost several thousand dollars if professionally done. It is better to shop around for one that already has a country-driving differential. If unsure insist on testing it at speed on a motorway.

Also gaining a following are small Hino coaches (Figure 4.9). They are are higher than Coasters but are within legal limits for use in Australia.

Avida and many others build motorhomes based on both delivery vehicles and medium and large truck chassis. The longer such units such as the 9.7 metre Avida Longreach (the forward part of the interior of which is shown in Figure 5.9) provide much the same space as a city home unit.

Figure 4.9. An interesting and affordable motorhome, is the 7 metre Hino coach. Pic. K.Allen.

There is an increasing move towards 7.5 metre motorhomes, often with two or three slide outs, as these are so much easier to drive, particularly in cities.

It also enables kerb-side parking in (Australian) urban areas - for which 7.5 metres is upper limit. Some motorhomes, however, are vast - and priced accordingly.

Coach conversions

Converted city single deck buses have low floors to ease access, but, as with many Toyota Coasters, most have gearing that is too low for comfortable long-distance travelling. An Australian-made overdrive solves that, but it is cheaper and more practical to start with a coach: these usually have gearing suited to cruising at 100 km/h.

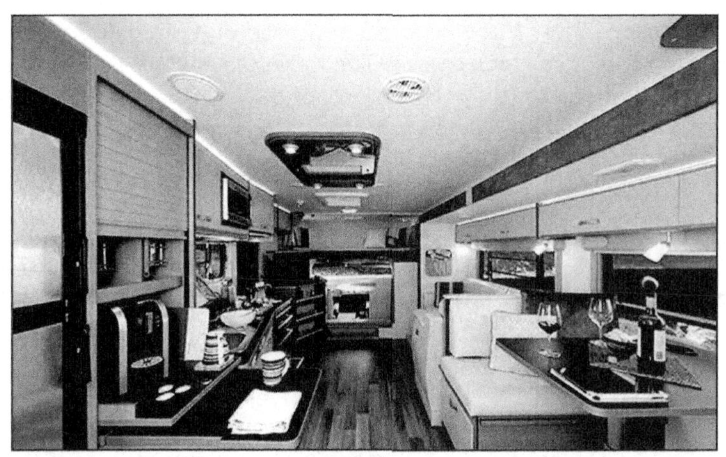

Figure 5.9. Avida Longreach motorhome (part) interior. Pic: Avida.

Most coaches have a high floor level with huge underfloor storage. They have similar rugged truck-like build but are often softer sprung and may have better sound insulation.

Given adequate ground clearance (and most have) coaches can readily travel main dirt highways. Their long overhangs, however, preclude them from tight creek crossings, and their weight tends to all-but exclude bush camping.

Converted coaches vary in price - from $20,000 to more than $750,000 and seemingly more dependent on condition and quality than age, (and, occasionally, reality!).

Fuel usage

Depending on speed and headwinds, a post 2012 turbo-diesel motorhome of about 4 tonne is likely to use 9-11 litres per 100 km, typically increasing by 1 litre per 100 km for every 1000 kg increase. Driving at 80 km/h instead of 100 km/h saves about 25%. This may increase by 30%-50% in sandy off-road going.

Figure 6.9. This lovely conversion is based on that Rolls-Royce of the coach world, the stainless-steel chassis MCI. Pic: rvbooks.com.au.

Chapter 10

Off-road campervans & motorhomes

The most essential requirement for driving off-road (including many dirt roads - that often have a high centre ridge) is adequate ground clearance. The minimum, typically under the rear differential, needs to be 200 mm - Figure 1.6. That, for the exhaust system, is preferably greater.

Figure 1.10. The need for ample ground clearance. Pic: Peter Wright.

It is also necessary to have adequate front and rear arrival and departure angles to avoid being grounded when climbing a sudden steep slope - as with narrow creek crossings. It is these clearances and angles (as much as traction), that can often determine where a vehicle can and cannot go.

Many owners with extensive off-road experience suggest that overall length should not exceed six metres. Anything longer is handicapped on ultra-tight bends. Width is best limited to 2 to 2.2 metres. Height should be no more than 2.4 metres as many tracks have low tree branches that are cropped at about that height to enable bush fire trucks access.

Laden weight for off-road use is ideally under 5.0 tonne. Anything heavier requires huge effort and a seriously heavy tow vehicle if seriously bogged (Figure 2.10). An off-road vehicle's allowable payload is usually far lower than its on-road equivalent so it can be hard to keep a laden 4WD motor-home within legal limits.

The OKA shown in Figures 1.6 and 2.6 weighs about 7 tonne (and many self-converted OKAs weigh much the same). They need not be that heavy. The author's previously owned OKA's interior was built from aluminium and spruce - enabling the vehicle to be only 5.2 tonne (Figure 5.37).

Figure 2.10. Overkill, but a heavy 4WD bogged in deep mud requires truly serious assistance. Pic: Peter Wright.

Need for four-wheel drive (4WD)

Four-wheel drive is no longer needed for any of Australia's major dirt roads. All are now closed by local authorities unless passable by 2WD vehicles. Four-wheel is, however, essential for desert crossings, the Canning Stock Route, the top of Cape York and the track into Jim Jim Falls in Kakadu National Park. It is obligatory for the track into Purnululu (Kimberley), and needed for gorges off the Gibb River Road, but not for that road itself.

For many owners, 4WD's main use is for accessing secluded camp sites down by a creek, and often more so for climbing out again.

A 4WD's typically low bottom gear also eases idling across rocks and creek beds but four-wheel-drive *alone* is not enough. If one wheel of a driven axle loses grip and spins, a standard differential cannot drive the other wheel.

This can be overcome by fitting manual or automatically locking hubs.Some 4WDs have them as standard. Auto-parts companies can advise.

Limitations of 4WD

First-time users often overestimate the traction provided by 4WD. If a heavy 4WD gets bogged, recovery needs knowledge and experience - and ideally another at least the same size to pull it out. The 4WD function should be used to assist getting out of trouble, not further into it.

Figure 3.10. Four-wheel drive cannot always help. Here the author is winching his 1940 QLR Bedford upright (in central Africa) after the ground gave way under its 7 tonne. Pic: rvbooks.com.au.

Figure 4.10. The Earthcruiser is available on a range of 4WD truck chassis. This one is on an Iveco. Most weigh a little under 4 tonne. Pic: earthcruiser.net.au

Campervan conversions

Most two-wheel drive campervans have small diameter wheels and tyres and very low ground clearance. A few have marginally more clearance but not remotely enough for less-maintained dirt tracks. Even if otherwise strong enough, few can be taken safely off-road.

Their 4WD variants are usually based on Japanese and European vehicles that are normally front-wheel drive, and engage rear wheel drive to assist on ice and snow. Some, however, are serious 4WDS. These include the Mitsubishi Canter, Isuzu, Nissan and Iveco.

Figure 5.10. The rare 4WD Toyota Coaster is effective off-road. This one is owned by Tidy Towns founder Glenys Anderson OA. Pic: rvbooks.com.au.

There are also 4WD variants of the Toyota Coaster. Some are factory made, but a number have been successfully converted by their owners and specialised companies. One successfully crossed the Simpson Desert accompanied only by a motorcycle.

Suspension

No matter whether two or four-wheel-drive, neither works unless the driven wheels are firmly on the ground and retain equal tyre loading. To ensure this, almost all serious off-road vehicles have a narrow beam-axle chassis that readily twists over undulating ground. So that the vehicle's body is not stressed as the chassis twists and flexes, such vehicles have the body held onto the chassis by three-point rubber or steel spring mounting. See Chapter 8.

Figure 6.10. This Bedford QL shows the chassis twist and axle movement possible with beam axle suspension. The QLR (Figure 3.10) had a similarly flexible chassis. Pic: Vauxhall/Bedford Motors.

Independent suspension provides a softer ride, but precludes tyres from having the equal loading provided by beam-axles. Further, chassis twisting (with independent front suspension) undermines steering geometry. For this reason it is rare to find a serious off-road vehicle with independent suspension.

Wheels & tyres

Tyres for soft going and dirt tracks etc. need to be of large diameter, but not width. Also avoid twin rear wheels and tyres because rocks become jammed between them, and on rocky going, one rear tyre (of a dual pair) may take the full load of both tyres. See Figure 12.11 in Chapter 11.

Avoid aggressive tyre tread patterns for all but use in rock quarries. Such tyres wear faster than road pattern tyres and also very noisy on bitumen surfaces. Dual purpose semi-road pattern tyres are superior in most conditions (particularly sand) and last far longer.

Expedition vehicles

These are primarily used to travel more or less globally but there is a reality chasm between some products offered and global off-road reality. Made mostly in Europe and USA, some 'expedition vehicles' are based on six or eight-wheel-drive chassis of huge size (and some weigh over 20 tonne).

Figure 7.10. Many dream of having a vehicle like this but, whilst fine in a desert or wide straight tracks, it is hugely too big to travel in many of Australia's and Africa's (etc.) most interesting remote areas. Pic: source unknown.

Whilst Australia's main dirt routes are travelled by road trains weighing over 150 tonne, once off such routes, even a 5 tonne, 6 metre 4WD vehicle can be a tight fit. Africa and much of India are similar. A further issue with such weight is that many bridges in third-world countries are rated at 7 tonne. Such mega-rigs may be fine for crossing deserts, but are far too large and heavy for travelling in many of the more interesting parts of the world. Finding camp sites for any vehicle is not easy. With mega-rigs it will often be close to impossible. so too is finding anything capable of aling There are also social considerations and risks if taking million dollar plus vehicles into poor areas.

Keep it simple

Construction and equipment should have the minimum of electronics. One well-known adventurer was plagued by obscure computer issues that rendered his vehicle immobile in Mongolia for two years.

The ideal (particularly for global travelling) was that arguably that designed and self-built for an Australia/Europe return trip (Figure 8.1). Aptly named the 'Tardis' it was based on a Canter chassis using ultra-light construction. The upper part telescopes down - enabling it to (just) fit into a standard size shipping container. It weighs an ideal 4995 kg. It was driven successfully overland across Asia to the UK and back.

Another excellent self-conversion is shown in Figure 9.10.

A vast amount of mostly useful information can be found on expedition-portal.com/

Figure 8.10. The Tardis: arguably the ideal vehicle for global travelling. For details see: epicycles.com

Figure 9.10. Australian writer, Jim Ditchfield's self-converted Isuzu - the perfect size for a go anywhere motorhome. Pic: rvbooks.com.au.

Chapter 11

Building your own

Building or fitting out a campervan or motorhome is much like owner-building a small house, plus knowing about electrics, mechanics, woodwork and solar. Only self-build if you really enjoy work like this as it may take several years to complete.

Many are built by those with no previous RV experience. Unless you have and know the priorities for the desired usage, layouts that seem fine on paper often prove less so in use, e.g, that seemingly essential (such as an in-built oven) is is often barely needed.

Establish needs and priorities by hiring an RV or two, and discussing your plans with experienced self-builder.s You must also ensure your base vehicle is appropriate. It is pointless to convert an OKA unless you really intend to travel extensively off-road: they are superb for that but huge overkill if not. Or convert a 12-14 tonne coach if you do intend to travel extensively off-road.

Some self-builders adapt to that made (or then modify the vehicle accordingly), but it is not uncommon to start designing the next one after a few months on the road! Ideally mock up a temporary layout using bits of old bedsteads, second-hand plywood, etc. and trial it some months before finalising.

Minimise weight: the lighter the vehicle the less fuel it consumes. Tyres, engine and transmission suffer less stress and wear, brakes work better. It will be less hard to extract if it becomes bogged.

It may pay to avoid using a diesel-engined vehicle as there are warnings that many countries will ban their sale (or even use) after 2030.

Legal issues

A National Code of Practice for Light Vehicle Construction and Modification and a National Code of Practice for Heavy Vehicle Construction sets out the requirements for modifying Light (under 4.5 tonne) and Heavy (over 4.5 tonne) trucks and trailers.

Australian Government
Department of Infrastructure, Transport, Regional Development and Local Government

National Code of Practice.

BUILDING SMALL TRAILERS.

INFORMATION FOR MANUFACTURERS AND IMPORTERS.

Summarised design and testing construction requirements for trailers that do not exceed 4.5 tonnes aggregate trailer mass.

Vehicle Standards Bulletin 1

Figure 1.11. VSB 1. This is currently (late 2019) being revised.

The Code sets out individual state and territory requirements for registration (excepting that some jurisdictions may have minor requirements over and above the Code). To ensure total compliance use your own state's or territory's version of VSB 06 for Light Vehicles and VSB 6 for Heavy Vehicles.

The dimensional limits for Light Vehicles are set out in VSI No 5, and for Heavy Vehicles in the Heavy Vehicle (Mass, Dimension and Loading) National Regulation at nhvr.gov.au

Also of value is VSB 1 (Building Small Trailers) and VSB 14 (National Code of Practice for Light Vehicle Construction and Modification). So is the Victorian Government's 'Conversion of Vehicles to Motorhomes' - (the previous requirement relating to vertical exhaust stacks no longer applies).

See also rvbooks.com.au/page/rv-road-rules-summaryaustralia.

Avoiding basic errors

By far the most common issues that prevent registration include: excess rear overhang (it must be whichever is the lesser of 3.7 metres or 60% of the wheelbase), over-width (maximum is 2.5 metres), over-height (maximum 4.3 metres), over-weight (the maximum is that on the vehicle's compliance plate), and non-compliant electrical wiring and LP gas installation.

For trailers that have one or more axles near their centre, the maximum permitted length from the tow ball to the centre of the axle (or axle group) is 8.5 metres. The rear overhang must not exceed whichever is the lesser the length of the front loading space, or 3.7 metres.

Other issues relating to trailers include: under-rated tow bars and tow balls, safety chain/s missing or inadequate, insufficient rear departure angle and/or ground clearance (the latter is a minimum 100 mm) and inadequate entrance/exit doors (one at least must be on the left-hand side or rear).

Weight

MDF 25 mm medium density fibreboard	18 kg/m^2
MDF 30 mm high density fibreboard	23 kg/m^2
Standard plywood - 25 mm	16 kg/m^2
Lightweight plywood - 25 mm	10 kg/m^2
MonoPan - 30 mm	5 kg/m^2
Steel sheet 1 mm (Colorbond/Zincaneal)	8 kg/m^2
Aluminium sheet 1 mm	2.7 kg/m^2
Water	1 kg/litre
Diesel	0.83 kg/litre
Petrol	0.73 kg/litre
Fuel/water tanks (including mountings)	approx. 15% of capacity
Batteries (deep-cycle)	23-25 100 Ah
Batteries (LiFePO4)	5-7 kg/100 Ah
Solar modules (makers claimed watts	3-10 kg/100 watts

Table 1.11. Weight of materials (approximate). Source: rvbooks.com.au.

It is vital to ensure the end result is well within the permitted maximum on road weight - most self-built RVs end up being far heavier than had been expected.

Also never assume that because an off-road vehicle chassis may have larger tyres and upgraded suspension it can carry a heavier load. Many have *less* allowable payload than their on-road equivalent.

If starting with an existing vehicle, strip out everything unneeded, then have the vehicle weighed before starting work. Record front and rear axle weights separately.

Before you start building establish the weight of all materials and inclusions. Much material commonly used is far heavier than generally realised (see Table 1.11 for examples).

The most common trap is to use fibreboard (MDF) for flooring, partitions, etc. The 30 mm version commonly used is 23 kg/m**2** . If used in a 7 metre caravan it weighs over 385 kg. It provides zero structural strength yet needs a heavier chassis and larger tyres to support it.

A braced 25 mm plywood floor would weigh 268 kg - It is strong but still adds unnecessary weight.

Better by far are composite materials like MonoPan. These are ultra-light yet have high structural strength. Those 25-30 mm thick typically weigh under 5 kg/m^2. They are amply strong enough for the floor and for frameless construction of the body shell. Walls require 25 mm, floors 30 mm. MonoPan can also be obtained in thinner sections for interior partitions and fittings, with light aluminium box or angle section used for framing.

Solar modules typically weigh 10 kg per 100 watts - and many RVs may require 600-800 watts. Some solar modules of similar size (but of marginally less output) now weigh under 3 kg/100 watts.

A typical (12 volt) 300 amp hour lead acid deep-cycle battery bank will weigh about 100 kg. Lithium (LiFePO4) batteries are only a third or so of the weight: the LiFePO4 equivalent will be 25-30 kg.

Applying such thinking across the whole project can save huge amounts of unnecessary weight.

Distributing the weight (campervans/motorhomes)

Most trucks and delivery-vans are designed to have their loading (apart from driver and passenger) distributed evenly fore-and-aft of the rear axle/s. Only a small part of the payload is intended to be carried by the front axle. Dual rear wheels and tyres (or dual rear axles) primarily carry that payload.

Figure 2.11. Rear (bed) slide out on a Sprinter campervan made by Chameleon Motorhomes (UK). Pic: Chameleon.

Where single rear wheels and tyres are used (e.g. as for 4WD trucks) their same size front tyres carry a lower load. They run at lower pressure accordingly. The recommended pressures are usually shown in the maker's specifications and/or instruction manual. If loading is unequal, the tyres on the lighter side must be run at the higher pressure required by the heavier side. See also Chapter 8.

Distribute the weight of heavy items such as batteries, tanks, major tools, etc. as low as possible, and as evenly as feasible across the loading area, and ideally with equal weight side to side.

Inadequately considered slide-outs complicate the above. Not just their own weight but, that of additional body and chassis bracing, can result in undesirable asymmetric tyre loading. Minimise this via equal and opposite weight, or better still, have a slide-out extending from the rear. This is also an issue with bed-over-cab RVs. With these overhead weight in this area must be kept to the bare minimum. Have nothing except light bedding stored there.

To ensure weight is correctly distributed, draw the side and top elevations and mark in the position and approximate weight of all loads.

For campervans and motorhomes the ratio of the distances from the effective centre of each load between the axles represents the static weight of the load on each axle.

Distributing the weight (caravans)

It is vital for on-road stability that conventional caravans (i.e. those towed via an overhung hitch) have a maintained nose weight (when laden) of about 10% total, and their rear end weight be as low as possible.

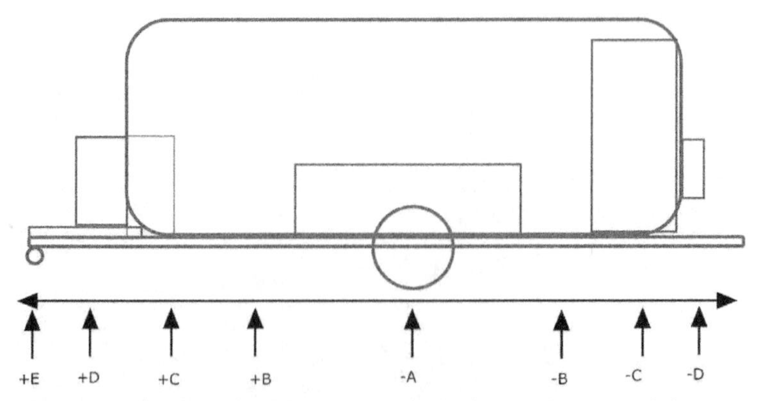

Figure 3.11. The effect of weight increases the further it is from the axle/s. Here, a 100 kg weight located 2.5 metres behind the axle acts as if it were 250 kg. This is why it is never a good idea to have that spare wheel at the extreme rear, nor heavy stuff on the A-frame (drawbar). Pic. rvbooks.com.au.

As the effect of weight along a caravan chassis is related to its distance from the axle/s it is vital to locate heavy weight as close as feasible above the axle/s. This effect is illustrated in Figure 3.11.

As tow vehicle makers progressively reduce permitted tow bar loading, increasing care is needed to reduce weight, and especially end-weight.

Have no heavy mass on the A-frame, nor spare wheels, tool boxes, etc. at the rear. The spare wheel should ideally be located underneath the caravan.

It is legal to have LP gas cylinders in a gas locker within the caravan. Australian Standard AS/NZS 5601:2.2010 sets out the requirements.

Caravan internal body length is ideally less than 6 metres. (The far-more stable fifth-wheeler format is preferable for caravans longer than this.) The ATM is best limited to under 1800 kg.

The axle/s should be located toward the rear such as to ensure a nose weight of about 10% of the total and ideally such that the nose weight does not vary more the 1% or so with varying water tank levels, nor any other form of loading.

Distributing the weight (fifth-wheel caravans)

The weight distribution and nose weight of fifth-wheel caravans intended for travelling (many are bought as non-mobile homes) is far less critical than for caravans towed via an overhung hitch. Nose weight can within reason be whatever the proposed tow vehicle can carry. There are only minor differences in stability for nose weights of 10%-25% of the trailer's overall laden weight.

Figure 4.11. The forward part of Glenn's trellis-type all aluminium 11.3 metre chassis. Pic: Glenn Portch.

Avoid stress concentrations by distributing the weight as evenly as feasible along the chassis, with the axle/s set back as far as possible such that the desired nose weight is achieved.

A good example all-alloy chassis of an ultra-light fifth-wheeler built by Glenn Portch. Glenn built eight of them - each progressively lighter per metre. The first, built in the mid-1990s, has since travelled over 300,000 km without any problems. Glenn's own 3200 kg, 11.3 metre Navigator is pictured in Figure 8.8.

Distributing the weight (camper trailers)

Camper-trailers too are often far heavier than expected: most have a laden weight of 1400-1800 kg and a few exceed two tonne (unladen). They do not have to be that heavy. The fully off-road TVan and the Ultimate, weigh well under 1000 kg.

Figure 5.11. Our previously owned OKA's rear. Its entire white powder-coated mainly aluminium interior designed by my (Scandinavian) wife Maarit, weighed under 75 kg. Pic: rvbooks.com.au.

Camper-trailers should have mass centralised but due to their short length (most have a body of only four or so metres) tow ball mass is less critical. Commercial products vary from 2.5% to 20%, but all seem equally stable.

Layouts

Commercial RVs give an idea of scale, but be aware that layout of many is dictated by the rental market. Such layouts typically provide sleeping accommodation for three or four people, at the expense of adequate storage.Study those of the specialised custom builders: some offer many variations.

With campervans and motorhomes, ensure there is direct access to and from the driving cab. Some use specially-made driver and passenger seats that swivel to face into a front lounge area.

Consider designing for light and space: our (pop-top) OKA layout (Figure 5.11) had no furnishings above bench level. Doing away with full-height everything above window level (except perhaps for shallow lockers above them), results in light and space. It is even possible with a shower/toilet.

Full-height wardrobes can be claustrophobic. Consider mostly having crease-resistant clothing rolled up in drawers.

Storage

The stuff you carry tends to expand to fill the space available. Arrange storage for what you need rather than the most you can fit in. Wire racks provide light, versatile, economic and space-effective storage. The baskets slide out for packing and or/use outside the vehicle. Dirt drops through the open mesh, easing cleaning by removing and hosing down the bottom basket.

Figure 6.11. Elfa wire baskets fitted into a powder-coated alloy cupboard. Pic: rvbooks.com.au.

Such racks vary in height, length and width and can be shortened if necessary. The baskets are available in various multiples of height. The Elfa brand is recommended for RV use. Cheaper ones tend to rust or fall apart.

Do not go to extremes: e.g. by providing exact space for items of which same-size replacements cannot be totally assured.

Utilise unused space. Jumper leads, hi-lift jacks, hoses, spares, etc. can be housed underfloor. Dust-proof hatches are stocked by boat chandlers.

Kitchens

First time builders tend to install kitchens far larger than later found necessary. The consequent space imbalance can then be difficult to remedy. Before planning, check out those in production RVs.

Consider having a duplicated outside cook-top and also hot and cold taps adjacent to the entry door. This keeps the interior cleaner and cooler in summer. The area needs some form of awning, and also a yellow light that does not attract insects.

Many experienced owners claim to rarely use an oven (other than a microwave) or, if they do, use a camp-oven over hot ashes. If you do install an oven you'll need an adequate air intake and an exhaust fan that vents to the exterior. The requirements for venting etc. are in LP gas standard AS/NZS 5601:2.

If feasible locate the oven such that, if unused, you can use that space for storage.

Installing a fridge

Refrigerators must be properly installed (Figure 7.11). Few are, and those that are not plague their owners ever after. Many fridges and fridge makers are then criticised because of incompetent installation, and often grossly sub-standard performance - over which they had no control. This is particularly so of three-way fridges. Those post 2000 work well but their correct installation is vital.

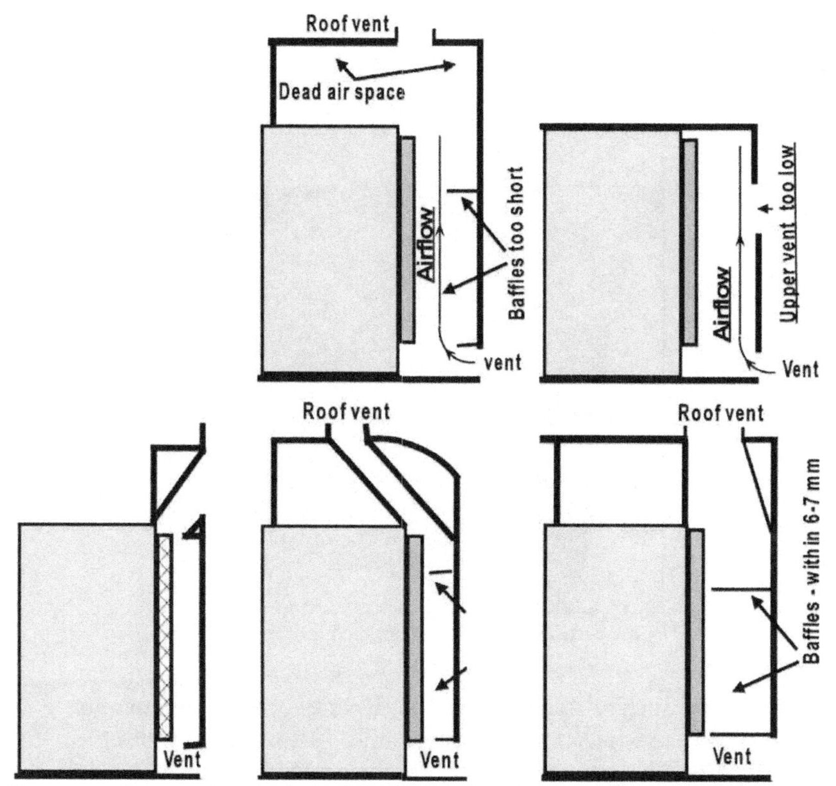

Figure 7.11. Top: too short (or lack of) baffles cause cool air to bypass cooling fins. Rising hot air is trapped in dead air space, hindering air through the vent. The upper vent is far too low: it must be above the upper fin level.
Bottom: how fridges should be installed. Pics: rvbooks.com.au.

Locate the fridge out of direct sun, with the wall behind it well heat-insulated. In all cases an intake fan run directly from a 5-10 watt solar module will further enhance performance (ex-computer fans are fine). Except for tropical use, no battery is required as cooling is mainly needed during the day.

Cool air *must* be able to enter at the fridge's base and be directed such that it flows over the cooling fins or heat-dissipating surfaces. It must also be able to flow to outside the RV with minimal restriction

If space permits, insulating a chest fridge with 100 mm of high quality material can halve its energy usage. Some fridges, however, dissipate heat only from their metal sides. These need a minimum 50 mm air gap to allow for cool air to flow up and over their sides.

Another approach is to make your own fridge. This is much easier to do than it may seem - not least as the compressor is remotely located. Boat owners often do this. Kits for doing this are available from Dometic, Engel and the (US) Sea Frost.

Insulation

Insulating the roof (solar modules assist) and walls to prevent condensation is essential, but avoid thermal mass, as that retains often unwanted heat at night.

Figure 8.11. Thermal batts are effective heat insulators.
Pic: source unknown.

Use insulation such as Astro-Foil. It has layers of polyfilm membrane between reflective foil a few millimetres thick yet works as well as 150 mm thermal batts. Fibreglass is also very thermally efficient, as too are shredded wool products.

Air-conditioning/heating

Air-conditioners made specifically for RVs are becoming more efficient, but some RVs use the latest smaller domestic units.

Figure 9.11.Daikin US reverse cycle air-conditioner is also an ultra-efficient heater. Pic: Daikin.

The Daikin US7 unit provides 2.5 kW (heat/cooling output) yet draws under 500 watts fully on and far less at lower settings. It can be used for short periods to provide 3.9 kW of cooling or 7.5 kW of heat. This unit is costly, but can be run from solar. It needs about six 130 watt solar modules plus 600 amp hours (or so) of lithium battery capacity if run all night.

Air-conditioning is becoming increasingly efficient but apart from the above and a few other units, most still needs mains power (or a generator).

Inside shower/toilet

Many RV builders see an inbuilt toilet and shower as desirable but to some extent the need is partially age and/or agility-related. If there's room for the toilet there is usually room for a shower (but they take up a lot of space). A good compromise is to install them in a commercially-available fold-up lightweight cubicle. An outside tent-located shower/toilet is also a workable compromise in small RVs.Chapter 16 covers toilets, their chemicals, water systems, etc.

Building off-road vehicles

A major issue when building an off-road vehicle is that this is a field where long-term development can be more important that the original design. The range of usage is so wide that it is impossible to simulate all possible conditions during initial design.

Follow the general approach of off-road products proven over time but use lightweight construction wherever applicable (such as MonoPan or similar) for the floor, exterior and internal fittings. Many expedition vehicles have been so made and proved totally successful.

The simplest off-road vehicles to build are camper-trailers. Apart from the hitch and suspension, these can be constructed virtually from scratch. But here too it is advisable to stay with existing designs.

Some camper-trailers have independent suspension but there is little point in having other than a beam axle that is properly located and sprung. Good shock absorbers are essential. The trailer should have at least the same ground clearance as its tow vehicle.

Campervan off-road conversion is tricky. Most have a delivery-van base (with off-road usage not remotely in mind). As many small commercial 4x4 trucks are only marginally bigger it makes far more sense to use one of those.

By and large off-road trucks withstand far more rough usage than do passenger-carrying 4WDs etc. Within reason their age is not an issue if they have been well-maintained and spares are still reasonably available. Despite that many have done over a million kilometres, most Australian-made OKAs now sell for a lot more than they cost brand new.

Off-road suspension

A beam-axled truck (whether leaf or coil sprung) with a chassis intended to flex is by far the best for off-road use. It enables huge wheel travel (Figure 6.10) with good weight transfer.

Figure 10.11. Unimog three-point body mounting allows major chassis flexing. Pic: original source unknown.

It is not feasible to have independent suspension with a flexible chassis. As one local converter found out the hard way, doing so seriously prejudices steering geometry, and hence handling and stability.

Beam-axle suspension is just fine, but rather than using the short stiff garden-trailer springs often used, consider using Toyota HiLux rear springs with a leaf or two removed, plus HiLux shock absorbers.

Accommodating chassis flexing

To accommodate the inevitable chassis flexing it is essential to have three-point pivoted and sprung body mounting. That used on the very flexible beam-axled Unimogs has long proven effective, but substantially increases their overall height.

It is inevitable, that no matter how sprung, some body flexing is inevitable. To avoid cracking or distortion, most body panelling is best held together by a marine-grade adhesive such as Sikaflex 252. This remains flexible yet grips so strongly you need a hammer and chisel to remove it. It is also ultra-strong - to the extent that OKAs' all steel body panels are held together solely by that adhesive.

Be aware that Sikaflex 252 adheres virtually instantly. If using it, position the panelling via pre-drilled holes, and locating studs that must be removed the instant the panel is positioned but not pressed down.

Dust exclusion

A major problem with many off-road vehicles is that, despite every opening being apparently sealed, dust tends to find its way in through even the tiniest of holes - particularly the sink plug hole (always have the sink plug in place whilst travelling) and worsened by body flexing.

Figure 1.11. A Donaldson Cyclone air filter: big and a bit clumsy - but they work. Pic: Donaldson.

Whilst every effort should be made to seal doors and windows etc, the only really successful solution is to have a high forward- facing scoop that forces air through a large air cleaner and into the truck. This maintains a higher internal air pressure - virtually precluding dust entry.

The big Donaldson units work very well for this but need cleaning every day or two in really dusty going. But that's far easier than cleaning out an entire RV's interior.

Off-road tyres

Extra-wide or rugged tyres may impress city dwellers, but outback owners usually opt for 750 x 16, or their 235 x 85 x 16 equivalent (both of dual-purpose tread pattern).

How tyre pressures affect footprint area

Narrow width enables the footprint to lengthen at low pressure - and hence act if it were a rubber caterpillar track. A wide tyre cross section 'bags out' at low pressure, but too high up to assist flotation. It also restricts footprint elongation.

Figure 12.11. Pics and data: Michelin Tyre Company and Peter Wright.

Electrical

An RV is a dangerous environment electrically. Safety requirements differ from domestic practice, (Chapter 25) and vary greatly between countries and also from time to time.

A major difference from domestic practice is that the neutral-earth (MEN) connection is not made within the RV. That link is via the supply cable and supply source. All circuit breakers and switches must be double pole. Not all electricians are aware of all this.

The legal requirements are set out in AS/NZS 3000:2018, and AS/NZS 3001: 2008, as Amended 2012. A Connection Certificate (from the electrician) is required for registration. Such certification is not required for systems below 120 volts dc.

Work out how and where to install wiring (and also water and gas piping), making allowances for future access. To save labour costs, commercially built RVs often have the wiring stapled to the framing between the inner and outer cladding - precluding later access without major work. Instead, run cabling through oversize conduit with access boxes (inside cupboards etc.) to ease subsequent changes or repairs.

Alternatively run it in conduit at floor level within the vehicle, bringing cabling up as required within cupboards etc.

Adequate cable is essential

Apart from often seriously-flawed initial installation, heavy cabling is essential for any RV electric fridge to run efficiently.

Even if the fridge is physically installed as shown in Figure 7.11 (and few are), if the cable is too light electric compressor fridges will still perform to a point, but will use excess energy (particularly in hot climates). Such fridges may be unable to maintain their specified cooling at very high ambient temperatures.

This, historically, has been a particular problem for a caravan-located battery that is powering a caravan fridge charged via a long cable from the vehicle's alternator. The only practical (and totally successful) solution is to install a dc-dc alternator charger adjacent to the battery in the caravan. It is also advisable to upgrade the cabling from the alternator and also, via the connecting cable and an Anderson plug and socket, all the way to that dc-dc unit. Using 13.5 mm^2 cable is far from overkill - but do not go below 10 mm^2.

Upgrading the cable transforms the performance of many fridges and fridge freezers. There is also a lot of information about this in the Articles section of rvbooks.com.au.

Chapter 12

Tyres

An annual survey of US RV tyre usage shows that one-third of all vehicles checked exceed their maximum loadings by 20%-30%. One noted that most owners grossly underestimate their loading. Similar checks in Australia and New Zealand indicate that, here too, many motorhomes, caravans and their tow vehicles' tyres are grossly under-inflated and/or overladen.

Index	kg	lbs
71	345	761
73	365	805
75	387	853
77	412	908
79	437	963
81	462	1019
83	487	1074
85	515	1135
87	545	1201
89	580	1279
91	615	1356
93	650	1433
95	690	1521
97	730	1609
99	775	1709
101	825	1819
103	875	1929
105	925	2039
107	975	2149
109	1030	2271

Table 1.12. The index number shows a single tyre's permissible maximum load.

Tyre ratings and pressures

Tyre ply rating has long since ceased to indicate the number of layers. It is now only a (vague) indicator of wall thickness and abrasion resistance. It is not related to loading capacity, but the higher the ply rating the greater its ability to run continuously at full load capacity. That load capacity is a function of tyre volume and pressure. It is shown on the side wall as an index number - see Table 1.12.

Ensuring correct tyre pressure is necessary for load bearing, and particularly for ensuring that the tyre's footprint on the road is of optimum shape and size. This can only be done by weighing each tyre's load. Where, as with some slide-outs, one side is heavier than the other, both sides' tyres must be set to the higher pressure required by that load.

Tyre makers' charts show recommended pressure for known tyre loading. Table 2.12 (later in this chapter) shows typical recommendations, but they may vary slightly from maker to maker.

Today's tyres are deliberately under-inflated when cold. They rise to their correct running pressure after a few kilometres. Set pressures only when tyres are cold. Do not reduce pressures as the tyres heat up: those days have long since gone. Some tyre fitters may advise otherwise if unaware that recommendations changed.

A few tow vehicles have dual rear wheels and tyres. As even loading of adjacent tyres cannot be assumed, each tyre of a dual pair needs 10% more pressure than would a single tyre. Dual tyres are rarely used for off-road vehicles because uneven loading occurs on rocky surfaces, and loose rocks get trapped between them and penetrate the tyres' walls. Dual tyres are also less effective in sand than singles.

Tubeless tyres

An inner tube constantly shuffles, generating heat and minor abrasion. Do not fit tubes to tubeless tyres. Friction caused as tubes move around increases the probability of failures. Carry an inner tube or two, however, as a tubeless tyre is hard to reinflate without a garage air compressor, but remove it when the tyre is repaired. Tubed Michelin tyres *must* have Michelin inner tubes or there will be ongoing flat tyre issues.

High pressure tyres

Some vehicles have tyres that require 830 kPa (120 psi) or more. This is way beyond the capacity of many garage compressors. If you have such tyres, carry a high pressure compressor. Wide-profile tyres are not good in sand. Tour operators may use them, but they usually have 14 or so passengers to push.

Tyre life

Tyre life is not just about wear. It depends on driving habits, storage time and conditions, weather conditions, loads and their distribution. All tyres oxidise internally, but flexing whilst driving 'pumps' the oxidisation to the surface, where it evaporates. Vehicles only driven occasionally have oxide trapped causing rubber and fabric deterioration. Tyres most at risk are old ones totally unused. Tyre makers advise that maximum tyre life is 7-10 years.

Figure 1.12. This tyre was made in the 13th week of the year 2001.

Tyres made since January 2000 have a small patch bearing a code sequence similar to that shown (Figure 1.12). The last four digits show the week and year of build. Here the code (1301) indicates a tyre made in the 13th week of 2001.

Tow vehicle and caravan tyres

Conventional caravans tend to snake (yaw). This imposes cyclic side forces on their tyres and those of the tow vehicle's tyres. Resisting this is aided by tyres with stiff side-walls but at the expense of minor increased road noise and ride comfort. As safety is important it pays to use the stiffer side-walled 6-ply light truck (LT) tyres.

Do not attempt to repair a tyre without removing it from its rim. It needs a plug inserted through the hole, and a reinforced patch on its inside. Repairs may only be made on the tread area between the shoulder grooves.

Removing and refitting tyres

Tyre beads tend to stick to the rims, and may need a tyre-removing device. Where one is not available, soak the bead with soapy water or detergent: it is usually then freed by jumping with your heels on the bead area. If that fails, push the wheel under the vehicle and, using a car jack, press down on a wooden block adjacent to the bead. Once free, push the bead *opposite* the tyre levers down and toward the wheel centre. This gives the clearance to lever it off. Tubed tyres require the tyre levers to be positioned either side of the valve.

Refit by mounting one side of the tyre onto the non-valve side of the wheel rim. Then, from the valve side, insert the valve (and tube if applicable). Push the bead directly opposite the valve firmly into the rim well (and if a tube is used) inflate very slightly. Lever the tyre gently into place on either side of the valve. Difficulties in refitting are usually due to the bead not being pushed fully into the well of the rim.

Tyre balance

Wheel/tyre balance is particularly important for large section wheels and tyres, and particularly with beam front axle vehicles. With those, tyre imbalance may cause the front axle to 'tramp': i.e. the wheels jump up and down and may even swing strongly from lock to lock. Another common cause of front wheel tramping is shock absorbers that are inadequate or worn out.

Tyres have a ring above the rim to show if they are centred. Before balancing, check that ring is concentric within 1-1.5 mm or it will be impossible to balance it correctly. With large tyres on small diameter rims (e.g. OKA's 900 x 16s) the heavy wheel drops out of centre if moved when the tyre is not inflated. To avoid this, inflate the tyre whilst it is lying flat before lifting it for balancing. Tyres that require more than three or four balance weights are usually faulty.

Table 2.12 shows typical loads and pressures. There are maker to maker variations particularly for smaller tyres: plus/minus 10% is not uncommon. The data in Table 2.12 is for single tyres. Each tyre of a dual-tyre set needs about 10% more pressure for the same load. This to allow for conditions where one tyre of a dual pair carries more load (as when traversing a ridge at an angle).

All tyres have a side wall indication of maximum operating pressure. The same size tyres may have various maximum ratings for load (as in the table on the previous page) and for pressure. There is also a rating for maximum speed, but all RV tyres are rated for a maximum that is above the speed limit in Australia.

An increasing number of new wide low-profile tyres have high minimum pressures of 550 kPa (or even more) that are required to maintain the correct profile. The very low profile tyres fitted to many SUVs vary a great deal from brand to brand. Their recommended pressures can be obtained from their makers.

Tyre Size	240 kPa	275 kPa	310 kPa	345 kPa	380 kPa	415 kPa	450 kPa	485 kPa	515 kPa	550 kPa
	35 psi	40 psi	45 psi	50 psi	55 psi	60 psi	65 psi	70 psi	75 psi	80 psi
205/75/15	670kg	710kg	750kg	790kg	830kg	870kg	900kg			
215/75/15	730kg	770kg	810kg	850kg	890kg	930kg	960kg			
225/75/15	800kg	855kg	920kg	975kg	1030kg	1080kg	1155kg			
215/85/16	695kg	760kg	790kg	800kg	845kg	950kg	965kg	1060kg	1130kg	1160kg
235/85/16	780kg	850kg	920kg	1000kg	1080kg	1155kg	1190kg	1255kg	1320kg	1380kg
255/85/16	895kg	980kg	1020kg	1120kg	1195kg	1240kg	1360kg	1415kg	1450kg	1550kg
750/16	735kg	805kg	875kg	925kg	995kg	1050kg	1120kg	1160kg	1210kg	1250kg
8.75/16.5						1020kg	1095kg	1120kg	1170kg	1220kg
225/75/17										
245/70/17	785kg	850kg	890kg	1000kg	1050kg	1090kg	1215kg	1240kg	1270kg	

Table 2.12 Typical data for tyres fitted to 4WDs and caravans.

Chapter 13

Lighting

Knowing what electric globe to buy used to be simple. The only choice was incandescent or fluorescent. As they all had much the same efficiency we used to buy them knowing that a 60 watt globe was fine (say) for the bathroom.

This has all changed. With the energy-efficient new technology available, light bulbs now produce the same amount of light using less watts. The better light emitting diodes (LEDs) now produce ten times more light for the amount of power than did incandescent globes: LED efficiency is now so high that it is pointless to use anything else. They cost far more initially, but typically last ten or more years, and use only a fraction of the electricity required previously.

We buy them now on the basis of the light they produce - using 'lumens' as a measure of the light produced. The higher the number of lumens they produce (per watt), the better the buy.

Figure 1.13. This 10 watt MR 16 LED produces 720 lumens and has a beam angle of 70°. Pic: Cree.

As a very rough guide a typical 60 watts incandescent light globe produced about 700 lumens. That amount of light is typically produced (in 2019) by one of the better brand 10 watt LEDs.

Light brightness and spread

A further and major difference is that, unlike incandescent globes, most LEDs produce a cone of light that is typically 30° to 120°. Whilst some light tends to be reflected if the walls and floor are light coloured, thereby providing background light in enclosed spaces, the light intensity on anything directly illuminated by an LED depends on the brightness of the light source (lux), the angle of the cone, and the distance from the light source.

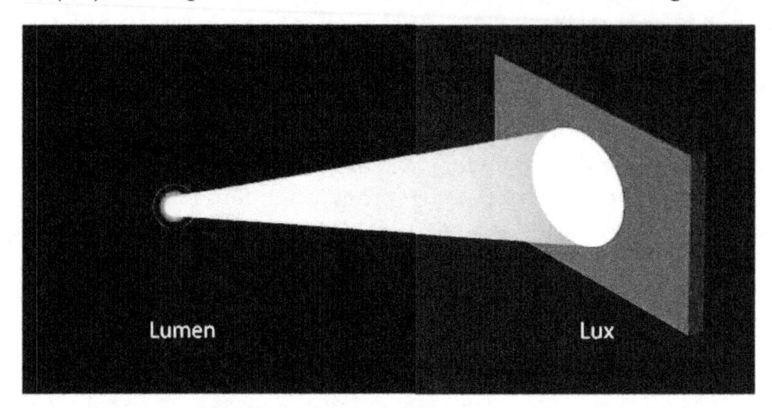

Figure 2.13. The relationship between lumens and lux. Pic: original source unknown.

The actual brightness of the light of that illuminated is measured in lux. The relationship between lumens and lux is illustrated in Figure 2.13.

Depending on the wall colours, living areas require background light levels of 50-150 lux. Tasks like reading and writing require about 350 lux. High precision work is likely to need 500-600 lux.

The actual amount of light required, particularly for reading and close-up work, is partially age-related. People (over 65 or so) usually need progressively more light as they become older, but as that is mainly needed for localised LED lighting, it makes little difference in energy usage.

Consult a doctor if a great deal more light is needed than noted above as it is often a sign that cataracts may need removing. This is now a relatively simple and painless procedure.

Colour temperature

More so than with other forms of lighting, LEDs are available in different forms of 'white' and often identified by their colour temperature (in degrees Kelvin).

Warm white (about 3100° K) is similar to incandescent and halogen bulbs. It is suitable for living and sleeping areas.

Cool white (4000° K) is a neutral light. It is often used for studies and kitchens.garages and workshops.

Daylight white (>5000° K) can appear harsh, unrelaxed and even sterile, but some people like it for bathrooms and laundries.

The move to LEDs

For RVs that have 12 volt halogen lighting, the existing light bases will be MR11 (for 10 and 20 watt) halogen, and MR16 for 35-50 watt halogens.

Figure 3.13. This GU10 type LED runs from a 230 volts supply.

Lighting suppliers understand these terms and can supply LEDs that fit directly into these bases. The MR type LEDs, however, have tiny pins that may not keep the LED in place on rough roads. Such fittings are best replaced by those that will. Here again, suppliers can advise.

A bonus of 12 volt LEDs is that they draw so little current that previously barely adequate wiring is now totally fine. They are also less voltage conscious: they do not flicker when the fridge turns on and off.

There are also many options for running LEDs from 230 volts. The compact GU10 is best for RVs. Its base is similar in size as MR16s but locks the LED firmly in place.

There are also 230 volt LEDs in both small and standard Edison screw form and now in the bayonet type light fittings.

TV interference

A number of ultra-cheap LEDs (and/or the tiny converters that drive them) are a serious cause of TV picture interference. Those known to cause this problem are eBay and chain store specials.

Light ratings and price

Light type	Approximate lumens per watt
Candle	0.3
Incandescent (including halogen)	5-25
Fluorescent tube	50-70 (80 for tri-phosphor type)
Compact fluorescent globe	45-70
Metal halide	60-115
White and warm white LED	120-1000 plus (in 2018)
White LED (prototypes)	1000 plus

Table 1.13. This table shows the relative efficiency of various light sources. It shows the relative brightness per watt as perceived by the average human eye.

LEDs vary over a wide range, as does quality. The cheapest are generally not worth purchasing. Most are very inefficient and some wreck TV reception.

The best buys (in lumens/watt) tend to be $12-$15. There are wide variations in price for the same products. The Phillips 7 watt GU10 varies from $14-$30.

Table 1.13. gives a rough idea of the respective efficiency of various light sources (as of 2019), but the best way to assess the light available is to use the output in lumens (a measure of the brightness of its source).

LED standards

Whilst LED lighting is an energy efficient alternative, official tests on some products showed a significant variation in product quality. Some lower quality LEDs produced little light, some flickered when dimmed, changed colour through life or failed prematurely. As a result, in April 2018, minimum standards for LED lamps in Australia and New Zealand (in line with European Union (EU) standards) are planned for September 2020.

Chapter 14

LP gas

In Australia, LP gas installation must accord with the Gas Installation Code AG 5601.2.2013 and be done by a certified gas fitter.

It is illegal to run an LP gas appliance from Autogas as that, in most parts of Australia, is a mix of butane and propane that may vary in proportion from batch to batch. Burning Autogas in appliances intended to run on LP gas can result in excess carbon monoxide. An accident so caused also invalidates insurance.

Portable gas stoves and lights use LP gas at cylinder pressure via tiny jets. They are cheap and simple but their jets can be impossible to clean, and hard to replace. Larger appliances have low-pressure jets supplied via a pressure-reducing regulator.

Cylinder location and ventilation

If the cylinder/s remain connected whilst the RV is moving, they must be in a fully open recess (e.g. within the profile of a trailer's A-frame and at least 200 mm above ground level) or in an LP gas locker subject to the requirements of the Code.

Figure 1.14. Dual LP gas cylinder housing. Pic: Glenn Portch.

Cylinders must be upright and held so that the fastenings withstand a steady load of four times the weight of the filled cylinder - with this load applied from any direction.

The cylinder compartment must have a drain of at least 25 mm diameter. Rigid conditions relate to direction, distance from any opening into the vehicle and source of drain outlet. Alternatively, there must be vents in the compartment or locker - each a minimum of 10,000 mm^2 of free area for every cylinder enclosed. The compartment must house the cylinder/s only, and allow easy removal and access to the cylinder valve.

Connecting to appliances

The LP gas regulator may be connected directly to the cylinder, and then to gas piping via a flexible hose (300 mm minimum length), or by looped tubing. It can alternatively be rigidly fixed and connected by flexible hose (minimum length 300 mm) and an excess flow valve; or by looped tubing. There are also various requirements for piping to appliances including its location, protection against abrasion and secure fixing.

Ventilation

Carbon monoxide gas is produced if there is insufficient oxygen for 100% combustion. There must be high and low level vents (at opposite ends) of a free area of 4000 mm^2 - or as below (whichever the larger):

V = (610 x U0 + 650 x p) here V = the minimum free area (in mm^2)

U = the input in MJ/h of all gas appliances including cook plates and stove (usually on the rating plate)

p = the number of people for whom that caravan (or compartment of a caravan) was designed.

For pop-top vehicles the ventilation must be equally effective whether the roof is up or down. For camper vans and motorhomes, no vent shall be installed on the rear end wall.

At least 50% of the related ventilation area must be no lower than 150 mm of the ceiling. If there is an externally vented range hood/extractor the vent must be no lower than 400 mm from the ceiling. At least 50% of the related ventilation must be as low as possible, but not more than 150 mm above the floor. The lower vent must have a sign such as: 'Never impede air flow through the obligatory vents.' Always turn off the gas at the cylinder whilst mobile, or if the vehicle is left unattended.

These vents ensure carbon monoxide remains at safe levels with correctly working appliances. Owners do install gas detectors but in small RVs they tend to respond every time one makes toast, or lights the oven.

WARNING

Ensure ventilation when the cooker is in use

Do **not** use for space heating

Figure 2.14. The lower air vent must have a sign like this.

Chapter 15

Water

Doctors recommend drinking two litres of water a day in cool conditions, two to four litres in warm conditions and four to six litres when it is very hot. More is needed if you are exposed to wind and sun.

It is important to maintain these levels. Not doing so can cause various medical conditions (including muscular cramping). Until recently it was held that beer, wine, tea, coffee, soft drinks, etc. are not substitutes for water and that they dehydrate the body. Experts now affirm that alcohol is still out but that tea and coffee in moderation is fine. Soft drinks are acceptable but water is best.

Water quantity

Drinking-water apart, the absolute minimum water consumption is around five litres per person/day. This is only just enough for cooking and cautious washing up, cleaning oneself via a flannel and a few centilitres of water plus a few drops more for brushing teeth.

Fifteen litres enables preliminary soaping plus a few seconds shower. Twenty litres is more comfortable. Add at least another five litres for each person/day when travelling long sparsely populated roads, e.g. between Adelaide and Darwin, or north of Cairns and Perth. Increase to 10 more litres per person/day if you are driving across long tracks such as the Tanami. If you do not need it, another may.

Obtaining water

Drinkable water is scarce across much of southern and western Australia away from the coast. Service stations usually let you fill water tanks if you buy fuel (but check first). Caravan parks will likewise if you are booked in. Public parks and gardens may have taps but the water is not always drinkable. Be wary of filling from taps in public toilets: there is a risk of faecal contamination because unthinking people rinse their cassette toilets under those taps.

Most outback service stations and mini-supermarkets sell 12-20 litre strong plastic containers of pure water at a very reasonable price. Many travellers use these routinely for drinking and cooking (often carried in the tow vehicle.)

If you intend travelling extensively in remote areas a separately installed pump and filter, and about 30 metres of flexible hose, enables pumping from streams and rivers. It also an intelligent way of carrying a spare pump. See also under 'Filtering' later in this chapter.

Water tanks

Small campervans typically carry (an inadequate) 30-50 litres, camper-trailers 50-100 litres, caravans and small motorhomes 80-200 litres, larger motorhomes up to 500 litres. Coaches may carry 1000 litres.

Standard-sized polythene and stainless steel water tanks are available from caravan and marine suppliers. Sheet metal fabricators will make up tanks to your specification. They usually cost much the same as mass-produced tanks, but it pays to shop around. Some manufacturers use aluminium, but stainless steel is safer for health. There are no problems with corrosion but some users report stress-related cracking.

Tanks should be housed under the vehicle and well above the lowest part. Caravan tanks should be as close as possible to the axle/s. If you intend travelling on dirt roads (and can accept the weight), tanks should be protected by 2-3 mm galvanised-steel sheet or similar thickness aluminium sheet. Remote reading water gauges show how much water is left.

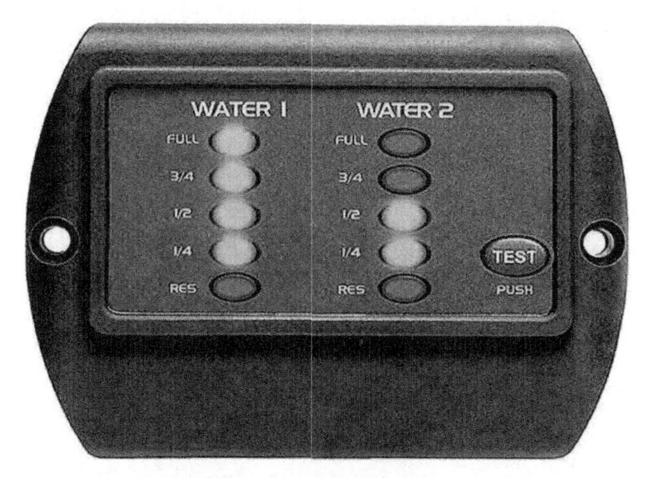

Figure 1.15. UK-made BEP water tank gauge. Pic: BEP.

There is a good case for two or more smaller tanks, rather than one large one, connected via 'Y-valves'. This assures drinkable water in the event of one tank being holed. Better still are separate systems - one for drinking water, the other for all else. You need a large diameter separate outlet pipe to atmosphere for air to enter when pumping, and escape whilst filling.

Slow filling and pipe burping is due to this outlet or hoses being too small or restricted. The breather outlet must be well above tank height or water is lost when crossing bumpy ground.

Water pumps

The simplest water supply is via gravity to a low-mounted tap, but manual or electric pumps are more convenient. Manual pumps are slower but are useful in that users draw only the minimum required.

Electric pumps deliver 8-12 litres per minute at pressures up to 280 kilopascals (40 psi), marginally lower than typical mains water pressure. Most draw 4-7 amps at 12 volts (48-84 watts) whilst doing so. The simplest systems have taps with inbuilt switches that actuate the pump. These are now rarely used as water and electricity are better kept apart.

Automatic pumps

For some decades, RV water pumps have operated by maintaining full pressure throughout the system. Whilst effective, the constantly changing pressure and water temperature is annoying, and night-time pumping disturbs sleep.

These issues are substantially fixable by installing a pressure tank (also known as an accumulator tank) that contains a strong balloon and is connected between pump and taps.

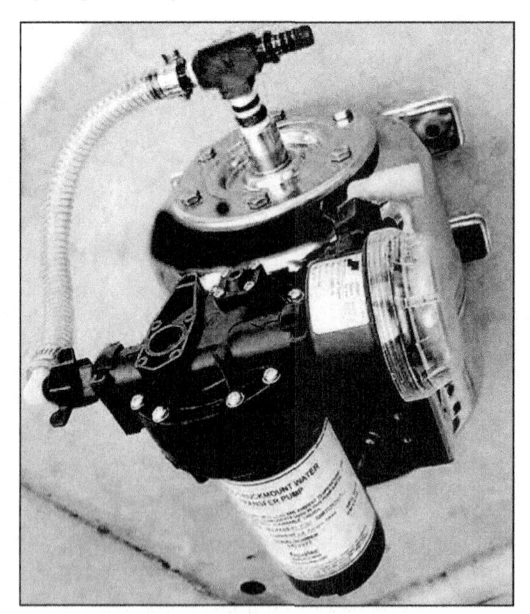

Figure 2.15. This system has the pump and pressure tank combined. Pic: Aquatec.

The balloon is initially inflated to about 100 kPa (19 psi). Water is then pumped in such that when the tank is half full, the balloon is compressed to 280 kPa (40 psi) or so. Water is then supplied by air pressure alone until it falls to about 100 kPa - when the pump restores pressure. This ensures that water is supplied.

Including a pressure tank saves energy because water pumps draw two to three times their running current each time they stop and start (and flexible pipes expand and contract with each pump stroke) further increasing energy use. They work best on vehicles able to accommodate a 100 litre or so such tank.

Figure 3.15. SHURFLO Smart Sensor pump matches speed to water volume required. Pic: SHURFLO.

There are also pumps that run at full pressure whenever a tap is turned on. Unwanted water volume is cycled around the back of the pump. This is effective but not energy efficient. Another, and more efficient innovation, is a pump controlled by a microprocessor that varies pumping speed to match the volume of water required.

Water pumps dislike long periods of non-use. Run them from time to time when the RV is not in use.

Quietening a noisy pump

Whilst they cannot be fully silenced, noisy water pump systems can be usefully quietened by arranging for flexible input and output hoses to be as loops that are free to move as the pump operates. It also assists to have the piping held within foam rubber wherever it is attached. This limits it transmitting noise to whatever holds it in place.

Water connection systems

Hose, and clamps made for reinforced hose are sold by hardware stores and irrigation suppliers. Spares can thus be limited to a few hose clips and a couple of metres of hose. Proprietary 'snap together' systems are neater and quicker to assemble, but complicates repairs in out of the way areas.

Mains water pressure in some areas is higher than the recommended 350 kPa (50 psi). It can be as high as 1000 kPa - about 140 psi. If you have a mains water pressure inlet, add a pressure reducer (they are available from irrigation suppliers).

Filtering

Distrust drinking water from anywhere except major towns. The main risk is giardia (faecal contamination) and cryptosporidium, but other less-hostile nasties can cause tummy upsets. Permanently installed in-line filtering removes most such, but not necessarily the undesirable heavy metals found in some bore water.

Avoid specialised housings that accept only their manufacturer's filters. There are no formal standards, but many makers have adopted a housing system for which interchangeable filters are available.

An adequate three-filter system has a readily accessible coarse screen filter between the tank and pump to remove grit and other mechanically damaging particles. Second and third filters, located between the pump and the taps, should be of 5-10 micron and 1 micron respectively. Filter makers claim that a 1 micron paper filter will trap the microscopic parasites giardia and cryptosporidium.

Some manufacturers supply various housings and filters for mains and low pressure water supply. Where there's a choice, use the low pressure version (but install the pressure reducer described above). Activated carbon and silver iodised filters etc. provide better filtering, but do not work effectively at low water pressure.

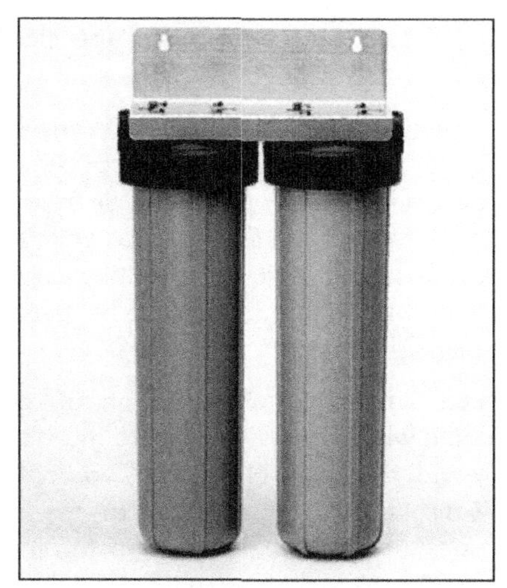

Figure 4.15. Typical dual water filters.
Pic: Big Blue-Twin.

Reverse osmosis systems produce excellent water, but use a lot of energy. They retain 10% or so of the amount processed - 90% is dumped. The residue could be used for washing etc. but it is not really worth the complication.

Self-containment - ('Leave No Trace')

Caravan park owners and some local authorities are increasingly upset by those seeking to free camp, but do not realise (or ignore) that many of today's larger RVs are substantially self-contained.

In 1998, the Campervan and Motorhome Club of Australia developed the 'Leave No Trace Self Containment Code of Conduct' to enable travellers to comply with environmental and ethical standards.

The Code asks participants to be aware of surroundings and only camp in appropriate places. Those who free camp should realise that whilst most people do not resent free-camping, tolerance may not extend to camping in inappropriate locations. Stuff strewn around, and laundry hung on trees is certain to annoy. This Code was later adopted by other Australian organisations. There is no universal 'code of requirements' but participants' vehicles must retain all waste, and leave no trace of their being on a site.

A typical requirement includes: at least 20 litres of fresh water, a grey water holding tank with a minimum capacity of 5 litres per person (or 15 litres per person if there is a shower), and a black water holding tank of not less a capacity than that of the smallest portable toilet cassette. Most campervans and motorhomes that comply with the above have tanks much larger than the minimums suggested: 100-200 litres is common. (Grey water must not be deposited on the ground unless specifically permitted by a controlling authority.)

Under the scheme, eligible vehicles may not stay in any Rest Area for more than five nights, or longer than the vehicle's self-containment allows. The Code suggests that participants pick up existing garbage in rest areas. The CMCA notes that 'this single act will do more for your welcome than almost any other'. It also exhorts participants to take care of the natural environment. If fires are permitted, they should be kept small and firewood collected only from appropriate areas.

Participation is open to all members of caravan and RV clubs who own a qualifying vehicle. People who wish to participate must sign a declaration that they will comply with the Leave No Trace Code of Conduct at all times. Qualifying RVs display their accreditation on the front windscreen, or for towed units, on the window closest to the entrance door and owner carries documentary proof of their participation.

Caravan and Motorhome Books thanks the CMCA for permission to reproduce parts of its 'Leave No Trace Self Containment Code of Conduct' scheme. See Chapter 38 for a Link to the CMCA.

Figure 5.15. The CMCA's Leave No Trace logo. Pic: CMCA.

Chapter 16

Toilets & showers

The smallest and simplest portable toilets have an upper section that contains a 15-20 litre water tank, a seat and cover, and a lower waste holding tank removable for emptying. Internal cassette toilets are similar but permanently installed. Some have an inbuilt water tank that needs periodic refilling.

Others plumb into the vehicle's cold water supply. Home-like toilets are also available that flush into a large capacity 'black water' tank. These need emptying less often, but when they do they must, as with all others, only be emptied into a suitable waste dumping point - see below.

For all such toilets, variously based chemicals are used to suppress odours, promote breakdown of solids, reduce gas formation, etc. This issue is covered in depth below.

Vacuum toilets

These work in a similar manner to aircraft toilets. When flushed, a strong vacuum flushes the contents into the holding tank. Only a small amount of water is required: the vacuum breaks down the contents. It is claimed that no chemicals are required. A downside is that these units are noisy when flushed.

Figure 1.16. Dometic vacuum toilet. Pic: Dometic Australia.

Alternative toilets

Alternative ways of handling and disposing of sewage are currently being developed. The most promising is the Norwegian-made Cinderella. This is a new approach that burns the waste product in an enclosed incinerator under the toilet seat. Combustion gases are expelled through a separate ventilation pipe.

Figure 2.16. The Cinderella LP gas toilet. Pic: siriuseco.

The RV version needs LP gas plus 12 volts dc for the ventilation fan and control module. No water or drain connections are required. Nor are you exposed to the waste.

The unit looks like a normal toilet but has a small bag (a new one is used each time) to catch the now ash-like burned waste. After use the lid is closed and a start button pressed. Combustion takes about 30 minutes. The burned waste needs emptying after 300 or so uses. This unit is now on sale in Australia. Information can be obtained at www.cinderellaeco.com

Toilet chemicals

The solid waste matter from toilets contains bacteria that, whilst beneficial when within one's body, are pathogenic to humans. Handling this is a major issue. There are currently two main approaches.

Bio-stimulant products are environmentally friendly. They speed up nature's aerobic breakdown of faecal matter by introducing oxidising agents or enzymes. They also reduce smells. Bio-stimulant treated waste can be safely disposed of in septic, environmental or city-type sewage disposal systems.

The second, highly undesirable approach (for disposal in septic tanks) method is a biocidic that kills *all* bacteria (both good and bad) indiscriminately. Any liquid so treated is allergenic and (claim some) possibly carcinogenic. It also requires a heavy chemical to reduce the stench of the hydrogen sulphide generated when excreta is broken down non-aerobically.

The major and serious drawback of any biocidic approach is that so-treated sewage can *only* be safely disposed of in city-type sewage treatment plants. As it kills all bacteria it also kills 'long drop,' environmental and septic systems. Because of this, caravan parks and other waste disposal facilities in rural areas are increasingly and rightly concerned about their sewerage facilities.

The best-known commercial safe products are Bio-Pak, Odour-B-Gone, Thetford's Aqua Kem Green and Aqua Kem Rinse (but not necessarily related products) and also BioMagic. Other products claimed to be bio-stimulant include EnviroChem, Reliance Bio-Blue, Century Clean N Fresh, and Chempace bioFORCE.

Absolutely (ecology-wise) it is essential not to use anything containing formaldehyde. The US Department of Toxic Substances also advises against deodourisers or other products containing Bronopol, Dowicil, or Glutaraldehyde for this or any toilet use.

Many long-term RV owners swear by Napisan (or Napisan look-alikes). These contain sodium percarbonate that breaks down into soda ash and hydrogen peroxide.

The labeling of the Napisan 'look-alikes' (Nappy Treatment Plus, Ultra Booster Everyday Plus Laundry Soaker, Laundry Soaker and In-wash Booster etc.) sold by the major supermarket chains, state clearly that these products are safe for septic tanks.

In a paper kindly provided to RV Books (formally Caravan & Motorhome Books) by Ian Jenkins the retired Professor of Chemistry and now Professor Emeritus at Griffith University states – 'In my opinion, sodium percarbonate is probably the cheapest, safest, and most effective product to use in portable toilets, provided it is used as directed'. The full paper can be accessed (free of charge) at https://rvbooks.com.au/page/napisan-a-professor-s-view/

Dumping points

Dumping points exist in most caravan parks, and many shire councils are installing them for visitors.

Never drain cassettes in public toilets and then rinse the container under the publicly-used water taps. These are used for drinking by unsuspecting people and such usage can result in seriously dangerous contamination.

Showers

Some (hardy) experienced travellers query the need for an internal shower on the grounds that all caravan parks provide them, and/or that in the bush an external shower is readily feasible. There is however a growing objection to soapy water being discharged on to the ground. It is, for example, forbidden in the CMCA's Bush Camping Code.

A shower cubicle's inclusion makes sense in interiors over five metres, but is hard to accommodate in RVs that are shorter than that.

It is, however, well worth considering carrying a shower tent (Figure 3.16). There are many on the market. It is readily possible to modify one such that soapy water is captured in a full width tray and disposed of in grey water systems.

Figure 3.16. A shower tent saves space, and many take only a minute or two to set up. Pic: source unknown.

Chapter 17

Space & water heating

Heating an RV space via LP gas is potentially dangerous because there is an exceptionally high risk of carbon monoxide poisoning. Carbon monoxide is a colourless and odourless gas produced when a carbon-based substance is burned without enough air - or if combustion is not 100%. It kills because the blood's haemoglobin, that transports oxygen throughout the body, is addicted to that gas. It picks up 250 parts of it for every one part of the vitally needed oxygen. There are only minor indications: headache, nausea, fatigue - then unconsciousness. If asleep at the time you are likely to die.

Safe space and water heating via LP gas necessitates a way that precludes carbon monoxide from that space. Despite this, advice such as 'invert a flowerpot or saucepan over an open gas ring', 'turn the oven on with the door open', and 'use a charcoal-burning cooking pot as a heater' is still posted on Internet forums. All are dangerous and all are illegal.

The Australian Gas Installation Code: states 'where a [gas] air heating appliance is installed in a confined space the circulating air shall be ducted and be separated from air for combustion and draught diverter dilution', as in the LP gas space heater mentioned below.

There are also restrictions on the positioning of LP gas storage water heaters etc. They must be installed by a licensed gas installer.

Claims that the catalytic heaters (once sold in Australia and New Zealand) are suitable for use in RVs, annexes and tents were withdrawn by the vendor many years ago. Because of the various issues and risks involved, diesel-fuelled space and water heating (and now also for cooking) - is substantially replacing LP gas.

Engine heat exchangers

Stemming from the 1920s, engine heat exchangers utilise otherwise wasted engine heat to provide limited hot water.

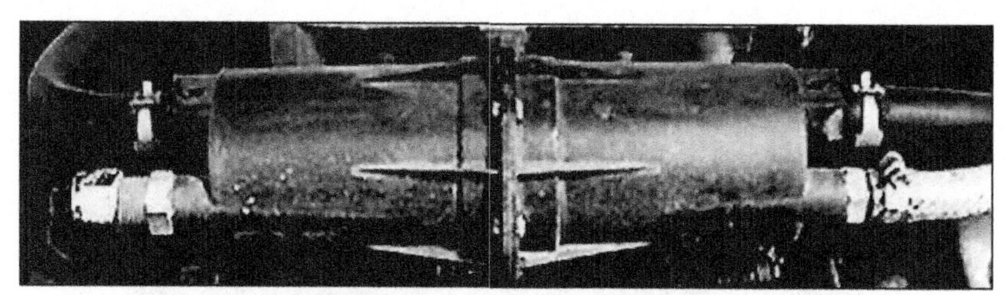

Figure 1.17. The compact Glind unit in-line heater is very effective whilst the RV's engine is at running temperature. Pic: Glind.

They have coiled metal tubing connected in line with the vehicle's interior heater. This tubing is inside a separate casing through which cold water is pumped - Figure 1.17. Heated water from the engine passes through the coiled tubing, transferring heat to the cold water. An external electric pump transfers the heated water to wherever it is needed.

Water temperature is varied by the vehicle heater's cold-warm setting, and/or by drawing more or less water through the heat exchanger via an adjustable restrictor. These heat exchangers work well enough but, unless the engine is running, are limited to providing a shower or two immediately after stopping.

Diesel/LP gas powered water/space heating

Diesel and LP gas powered space and space/water heaters have been used in boats since the 1930s, and in vehicles in Europe for many years. They are now becoming increasingly used in RVs worldwide. The main suppliers are Webasto, Dometic (Eberspacher), Diesel Heating Australia (Genesis and Snugger), and Truma. These units draw outside air into a small sealed furnace that burns injected fuel and exhausts the 'burnt' air to the atmosphere.

Figure 2.17. The Truma E 2400 LP gas space heater. Pic: Truma.

The furnace is within a sealed metal enclosure and heats the air blown between the two. As the external air and exhaust are routed to the outside, burning gas is sealed from the air (and/or water) heated within the vehicle.

Truma's E 2400 space heater has a rated thermal output of 2400 watts. It uses 100-200 grammes of LP gas an hour.

The Dometic and Webasto diesel equivalents have much the same heat output. They are claimed to use 0.1-0.3 litres of diesel fuel an hour.

Combined water and space heaters (Figure 3.17) work much in the same way, but heat a glycol-based fluid pumped through the outer jacket of the heater. This fluid is heated to a high temperature and circulated through an external calorifier (heat exchanger).

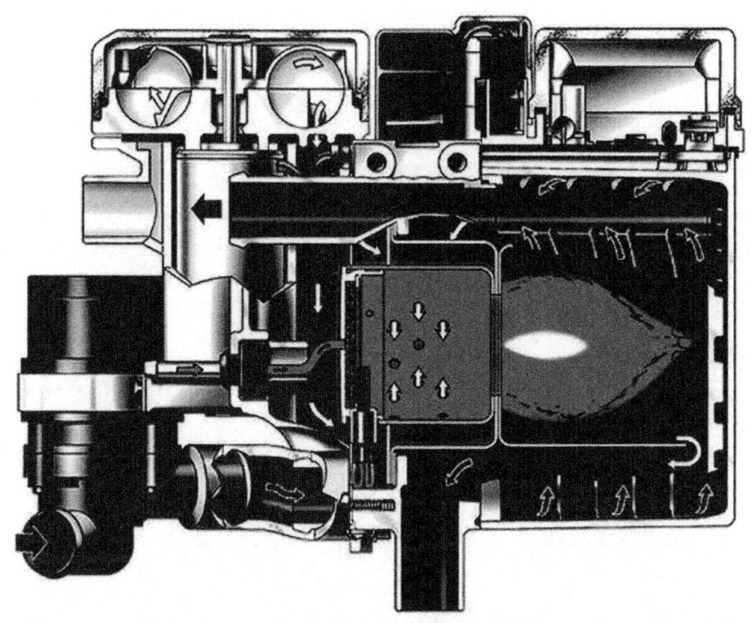

Figure 3.17. Webasto space/water heater. Outside air is burned together with injected diesel in the furnace (right). Water/glycol mix enters (bottom left). It is heated as it passes over the furnace and is ducted to an external heat exchanger via the orifice (upper left). The Dometic unit (made by Eberspacher) is generally similar. Pic: Dometic Australia.

The water to be heated is pumped through the calorifier (that also stores the heated water) to hot water taps after an initial few minutes. In some, the heated water reaches well over 70°C. To safeguard against scalding, a tempering valve (set to 50°C) that dilutes the hot water with cold water is legally required in Australia. It is strongly recommended to be used everywhere.

Wth some units, heating is optionally provided by the hot glycol being pumped through one or more fan-actuated radiators. A small control module can be located wherever convenient.

Fuel for diesel versions is supplied by a tiny electric pump and filter from an (optional) 10 litre tank, or via a tapped connector from the vehicle's diesel tank. The latter usually ensures cleaner and fresher fuel.

Stale diesel may build up fungi. This can be prevented by adding a diesel anti-fungi additive - sold by most truck fuel stations. Using an anti-fungi additive is advisable for anything diesel-powered that is unused for more than a month or two.

Exhaust and inlet silencers quieten the unit such that it cannot be heard more than a few metres away.

Most makers offer two or more sizes of heater. The smaller/smallest is adequate for the average sized RV and comfortably heats the interior and annexe of a caravan or motorhome on the coldest of nights.

The heat output of the air heating (only) units is too great for some camper-trailers even on the lowest setting. That can be overcome by it also heating an annexe. A better way, however, is to use a combined space/water unit as the interior-heating part has a wider range of control.

Experience with the smaller (Webasto) air heating version in our previously owned OKA (interior length 5.2 metres) was that even on a low heat setting, the heater kept the interior at a comfortable 25°C even when freezing outside.

The combined (space/water heating) unit in our later owned TVan (Figure 4.17) provided ample hot water for three weeks on less than 3 litres of diesel.

Figure 4.17. Webasto dual space and water heater fitted neatly into the modified front of the author's TVan. Pic: rvbooks.com.au

Chapter 18

Refrigerators (fridges)

Refrigerators pump heat from their insides (where it is not wanted) to where it does not matter. Their ability to do so, and the power they draw, increases with their surface area. Because of this (as a quick sum or two will show) one large fridge will generally use less energy than two small ones of the same combined volume. Likewise a tall thin fridge will use more energy than one closer to square. Chest opening fridges are thus more efficient, and their contents thrash around less whilst driving. Their downside is that items most needed somehow migrate to the bottom.

Because hot air rises and cold air falls a door-opening fridge is slightly less efficient that a top-opening fridge, but that can be reduced by adding plastic high-fronted drawers. The door seals must be in good condition as energy usage soars if they leak.

Electric fridges

Until recently, most conventional electric fridges drew power until they reached their preset temperature. Power was then cut until the internal temperature rose 1°C to 2°C above that. They thus cycled continuously with the ratio of on/off time varying with the preset and ambient temperatures. On really hot days, a marginally performing fridge could remain 'on' all the time. Cooling and energy use suffered accordingly.

Whilst simple and effective, on/off cycling is inherently inefficient, not least because a fridge's compressor motor draws several times its running current for a second or so each time it cycles on. To remedy this, there is an increasing trend to use continuously-running compressor motors that maintain the set temperature by varying their speed. This has proven to be a typical 25% more energy efficient.

Regardless of the cooling technology, for virtually any type of fridge energy usage increases by about 5% for every 1°C above a rated 25°C ambient temperature. It also increases by the same amount for every 1°C the thermostat is set below 4°C. Energy usage may thus increase by half in northern Australia, and double or more if a fridge/freezer is used to cool and store freshly caught fish.

Fridges made specifically for RVs typically draw about 8-10 watt hours a day per litre, progressively dropping to about 6 watt hours a day per litre for units over 180 litres. There has been some improvement in energy efficiency, but not so far comparable to that of 230 volt domestic units.

Of domestic fridges, there is an increasing difference in energy usage between the most and the least efficient. The best (such as LP and Miele) are 230 volt units of 200 litres upward. Some use only 50% of the energy of earlier ones. These work well in the larger RVs if run from one of the latest inverters - some of which are over 96% efficient (Chapter 27).

Consumption is affected by the thermal quality of a fridge's insulation, how well the fridge is installed (Figure 7.11 in Chapter 11) and individual usage. The energy draw of most fridges can be slashed by adding more insulation.

Three-way units

Three-way fridges run from the vehicle alternator whilst driving, 230 volts ac when available and LP gas at all other times. Their electrical consumption is far too high for running from solar alone.

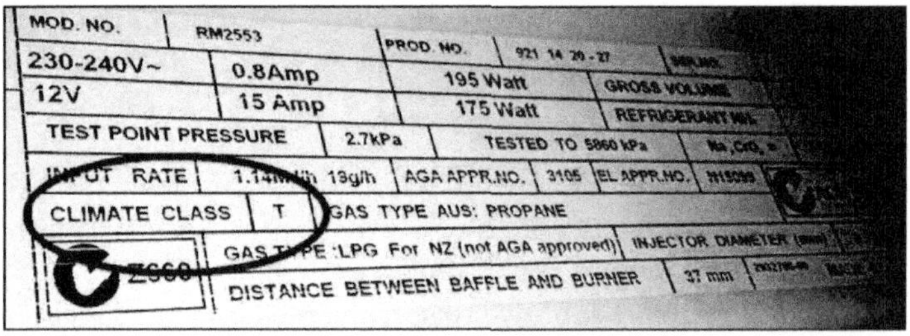

Figure 1.18. A three-way fridge's Climate Class is shown on a label inside the fridge. If it is not labelled as shown here it is not a Climate Class rated fridge. This is a 'T-rated' unit. Pic: source unknown.

Such fridges still have a long-since unwarranted reputation for being ineffective in hot climates. This is primarily because there were no recognised associated Standards until year-2000, and also due to often incompetent installation.

These fridges now have four tightly defined (CEN Standard) Climate Classes- that relate to ambient temperature: 'SN' and 'N' (Sub Normal, and Normal) work up to 32°C, 'ST' (Sub Tropical) up to 36°C. and 'T' (Tropical) up to 43°C.

One drawback is that T-rated fridges appear not to work that well in ambient temperatures below 18°C.

It is not easy to establish Climate Class from the promotional literature, but it is shown by the letter to the right of 'CLIMATE CLASS' on the rating/compliance plate inside the fridge - Figure 1.18.

Do not simply accept reseller's claims re this. They may not intend to mislead but many queried in the process of rewriting this book were *still* unaware of the Climate Class concept - despite it being used since the late 1990s. Some truly (but wrongly) believed that Dometic's use of the term 'Tropicalised' for all of its post-2000 product implied *all* were T-rated'. It does not, and Dometic has never claimed they were.

A correctly installed 'Climate Class T' fridge will work satisfactorily in tropical areas, but less so in cold areas below about 14°C. Three-way fridges are costly and LP gas is an ongoing expense, but as electric fridges (in the smaller RVs) typically draw 70% plus of all electricity used in that RV, their high cost is offset by the far smaller solar and battery system then required. There are also no concerns about long overcast or rainy periods.

Eutectic fridges

Eutectic type fridges (e.g. Autofridge, Indel) are usually 12 volt dc and 230 volts ac. They have internal holding plates that contain a low freezing point mixture - such as water and alcohol.

Figure 2.18. The Italian-made Indel eutectic units have a loyal following. Pic: Indel.

When first switched on, the thermostat is set to maximum cold and the fridge is 'pumped down' for 8-10 hours. Thereafter it needs power for only an hour or two each morning and evening. Such fridges can also be used (without prior pumping down) on their thermostat setting, but then use more energy.

The major advantage of these fridges was previously that (in pump-down mode) their compressor motor ran continuously, i.e. they did not suffer from excess power being drawn by the motor's constantly stopping and starting.

This benefit ceased when fridges began to use variable-speed compressors. A eutectic's main benefit has become that it need not be run at night - a boon for light sleepers.

The Australian-made Autofridge is chest opening. Indel makes a eutectic door-opening fridge. Victoria's Ozefridge produces eutectic systems intended for building into existing fridge cabinets, as does the USA's Sea Frost company.

Running on solar

The more efficient electric-only fridges can, if required, be run from solar if used for normal cooling and freezing. As a rough guide, 220-250 watts of solar and 100 amp hours of battery capacity is needed for each 100 litres of fridge capacity.

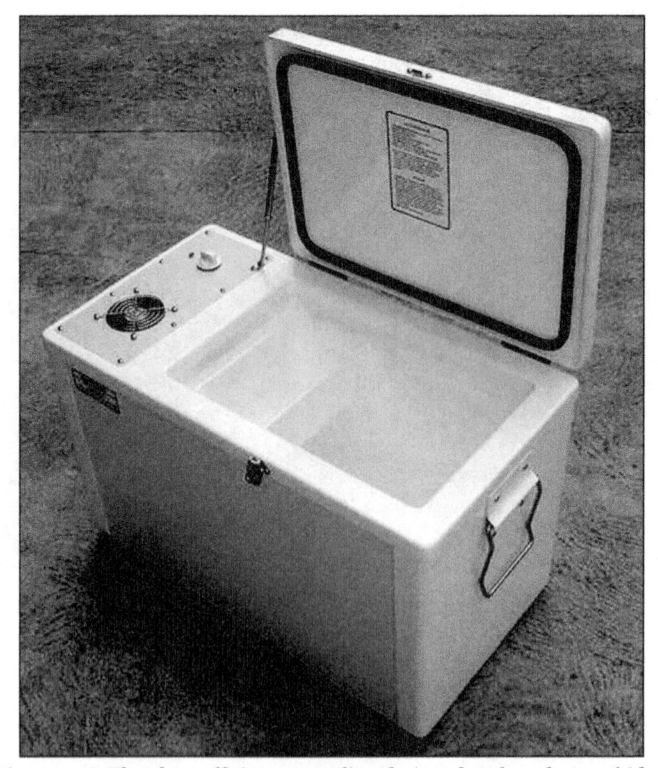

Figure 3.18. The ultra-efficient Australian designed and made Autofridge. Pic: Autofridge.

It is not feasible to use solar alone for cooling and freezing large quantities of fish. The energy needed is much too high. The only current solution is a quiet generator.

Fridge problems

RV fridges made since the late 1990s are reasonably efficient and reliable, but many users find their own is not. Fridges do occasionally need repairing (and three-way fridges need servicing every five years or so) but almost all fridge problems are due to inadequate or downright incompetent installation.

This particularly applies to three-way fridges - to the extent that many RV owners, unless fully aware of the above, swear they will never buy another one.

Making a few simple changes can transform many a fridge's performance. Figure 7.11 and Chapter 11) shows how.

Chapter 19

Television

Australia's analogue TV service ceased in 2014. The now digital system reflects the quality inherent in the program material. There is no snow, ghosting or sound distortion. You either have a picture as perfect as your TV allows, or it is non-watchable.

A digital signal converter can still be added to an analogue TV but, as post-2014 TVs use far less energy, it makes every sense to buy one that's digital.

Figure 1.19. CL 50 cm (23 inch) TV is ideal for small RV use. Pic: Harvey Norman.

A major change was the Australian launch of Freeview - a service that unites all free-to-air broadcasters so as to better compete against subscription television (in particular Foxtel). It is a worldwide concept with close to 25 countries having already adopted the standard.

The services, that in effect began in mid-2009, now include an enhanced electronic program guide. They also enable the use of TV set-top boxes and personal video recorders (PVRs) that meet the technical requirements.

In June 2010, an enhanced version (Freeview EPG) devices became available in retail stores. These versions have ceased operating.Freeview now focuses on its Hybrid Broadcast Broadband (HbbTV) TV-based FreeviewPlus service.

This initiative establishes a standard ETSI TS 102 796 (1) that in effect enables the reception of digital television from sources such as traditional broadcast TV, Internet, etc. via satellite, cable and terrestrial networks, etc.

It accepts content from an external hard drive (and cloud storage), and also from advanced interactive services and Internet applications.

Whilst the above can be dauntingly technical, all buyers really need to ensure is that the product is HbbTV compatible. If so, it is likely to have a FreeviewPlus logo on it. There are various versions and in varying colours - but most are like that shown in Figure 2.19.

Figure 2.19. The FreeviewPlus logo varies but most are like this.

HbbTV can be built into any internet-connectable TV, video recorder or set-top box. Manufacturers are free to include HbbTV but if they wish to use the FreeviewPlus logo, however, they must abide by Freeview's rules regarding the ability to skip advertisements, fast-forward, etc. Some TVs, however, are HbbTV compatible but their makers do not bother to obtain certification.

Screen types

LCD/LED TVs draw far less power than do plasma units. A typical 80 cm (32 inch) unit now draws 40-50 watts, a 2019 36 cm (14 inch) TV may draw as little as 10 watts. HD (high definition) units draw more. Energy draw is also related to picture brightness and definition. It is also feasible to use a USB HD TV adaptor and a laptop computer.

Recording

Video cassette recorders were replaced by DVD recorders that in turn morphed into personal video recorders (PVRs). Recording then became (comparatively) easy via electronic program guides automatically received and displayed by the PVR. These guides also précis program content.

To record onto the inbuilt hard disc or USB, scroll through the guide, select the desired program/s, and press 'Record'. Time and date, etc. are set up automatically by the unit. Your recorded programs are listed by name and content précis. Some allow you to 'bookmark' where you are up to in a program - handy if you are hit by a sleep wave!

Many TVs have a USB slot but not all enable the TV to record. Apart from attempting to read the lengthy associated manual, look for a 'record' button. It is usually red, or has a red dot. Some TV sets accept a USB connected hard drive, but most require a powered USB drive for storage sizes exceeding 250 GB.

Tuning

With current digital TVs, you usually select 'Installation' and then 'Search'. The TV does the rest. Some early set-top boxes retain all scans, resulting in confusing listings. In this event, use the factory reset (see instructions) before re-scanning. Today's set-top boxes do this automatically.

Digital TVs have 'logical channels'. To select, simply press the station number. The 'code' stays the same across the whole of Australia, e.g, for the ABC press 2, for SBS press 3. (If high definition press 20 and 30 respectively.

Existing antennas that satisfactorily receive analogue signals will handle digital signals. Directionality is often less vital in cities, but is very much so in low signal areas. Aligning the antenna is eased by using a signal strength meter. To reduce picked up electrical noise - use only RG6 quad-insulated antenna cable.

An optional antenna booster may reduce signal losses. Most are powered via the signal cable but a separate power feed is better. Too strong a signal can be as bad as too weak. The Strong (brand) unit responds to overly strong signals, but for most boosters it is necessary to assess this, and switch them off.

If feasible turn the TV off (at the wall switch) when not in use. It otherwise continues to draw power.

Satellite TV

Satellite TV requires a dish antenna that has a clear line of sight to the satellite (for Australia that's above Samoa). Trees block the signal. Some RVs have both dish and normal antennas.

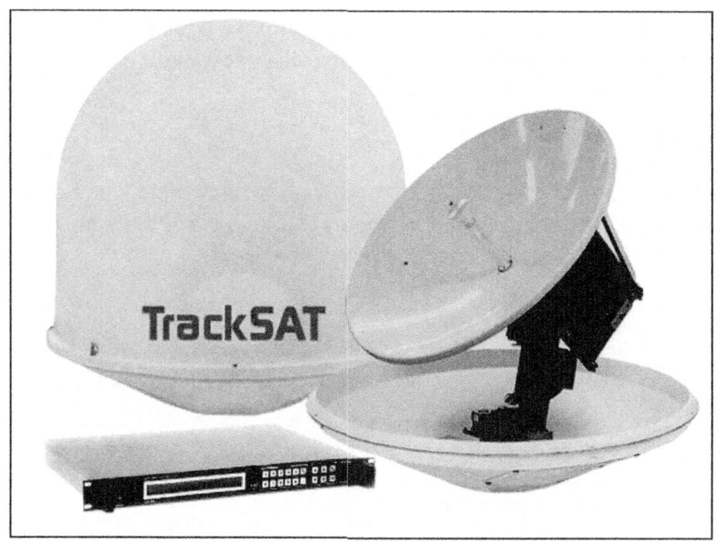

Figure 3.19. TrackSAT's UltraTrack UT100 uses servo stabilisation and inbuilt GPS to lock onto signals. Pic: TrackSAT.

A basic service needs a decoder, an LNB (Low Noise Block), connecting cable, an inclinometer and a compass. Initial tuning really needs someone to show you how, and to tune the TV. From thereon setting up takes only a few minutes. A 100% result is unnecessary - minor errors in tuning make little discernible difference in performance. Automatic satellite systems save time but cost around $5000.

You may legally access the ABC's and SBS's extra channels including High Definition.

One bonus is 'time shift': miss the 7 pm news in NSW and you can access it from the Queensland stations (one hour behind) or WA's (three hours behind).

For current information regarding free viewing see freeview.com.au/about/

Digital radio & satellite radio

Digital radio & satellite radio are widely available. The ABC and SBS operate digital radio services. Most are national, the ABC provides the local ABC Local Radio stations for respective locations. The local radio services are mostly a simulcast of their AM radio equivalents. The ABC has various digital-only radio stations, including ABC Jazz, Double J, ABC Country and ABC Extra - that covers special events. See digitalradioplus.com.au/listen

The also-available satellite radio reception provides perfect reception where there is no terrestrial coverage (e.g. most of outback Australia during daylight). A huge benefit of satellite radio for travellers is the ability to listen to news etc. from their home states, and particularly to obtain advance local notice of cyclones across Australia's top end.

Caution

TV technology, marketing and programming changes rapidly and it is inevitable that some of the above will change - even over a short time. Readers are advised to check before making buying decisions.

Chapter 20

Communications

Telstra Next G, 3G, 4G and now 5G provides mobile phone and wireless broadband across most of Australia's population areas and sections of some main highways. In late-2019 it catered for over 96% of the population. Outback coverage, however, still extends only a few kilometres from the transmitters. For full outback and remote coverage the only solutions are HF radio (now less used) and satellite telephone.

Email

Email can be accessed in some public libraries, tourist information centres, Internet centres and cafes - or via your personal computer or smartphone. Mobile use requires an appropriate modem and an email account. Text messages and email too can be sent via Next G and similar mobile telephones.

Satellite telephony

Satellite telephones work to and from anywhere worldwide except (late-2019) North Korea. Once the size of a suitcase, satellite hand sets are now only marginally larger than cell phones. All need unobstructed line of site to the satellite. For use indoors they may need an external antenna. Mobile calls require dialling only the mobile's number. Calls to land line numbers require dialling full international and country access codes.

Figure 1.20. Satellite telephone. Pic: Isatphone.

The Vodafone satellite unit first checks if a land-based network is available. If it is, the unit automatically selects it. If not, it switches to satellite operation.

Sat phones are not chat phones: these telephones are still costly - with a fixed charge of about $30 a month plus usage charges. Calls to and from some countries are up to $10 minute.

HF radio

High Frequency (HF) radios have up to 500 channels, but traffic congestion is high. Connecting to the public network via HF is still available but (except for emergency calls) it is no longer possible to make or receive personal calls via the Royal Flying Doctor Service. HF radio handles limited data, but only to another data-equipped HF receiver.

The HF radio system served outback Australia well but (apart from amateur radio use) satellite communication is now by far the better choice. Such telephones are readily available for short-term hire.

CB radio

This is handy for chatting or exchanging information, and for vehicles travelling in convoy. Single Sideband CB 27 MHz sets extend usable distance, but not reliably nor predictably.

UHF units operating at 477 MHz provide the quietest, high quality transmission. Their range is mostly line-of-sight but, where available, UHF repeaters can extend that to 300 km or more. Outback stations use this service extensively.

Pocket-sized UHF transceivers provide line-of-sight communications. They are handy for communicating to and from the vehicle (and also to enable a passenger to assist the driver when reversing into tight areas).

Channel usage

Whilst CB radio is not reliable for emergencies, channel 9 (27 MHz) and channels 5 & 35 (UHF) are reserved for emergency use.

Channel 10 is intended for 4WD drivers, convoys, clubs and national parks. Channel 18 is intended for caravanners and campers, but, as CB enthusiasts use that for chatting, some resent 'others' using it.

The Campervan and Motorhome Club of Australia suggests using channel 20 (both 27 MHz and UHF).

Channel 29 is used for Pacific Highway (NSW) and Bruce Highway (Qld) road traffic.

Truck drivers mostly use Channel 40 (UHF). It can be entertaining but carries language that parents naively believe their children do not use outside their hearing.

For changes see: uhfcb.com.au - particularly regarding usage of channels 41-80.

Chapter 21

Away from mains power

A vehicle's alternator supplies electrical energy whilst the engine is running. Its associated voltage regulator ensures that sufficient energy is available at the correct voltage for on-road needs. This includes replenishing the starter battery that, at present at least, can be used to charge the auxiliary batteries used for the RV's 'house' system when away from mains power.

Solar began to be included around 1985 and became increasingly cheaper and efficient. It typically charges 12 volt or 24 volt batteries. Whilst simple, 12/24 volts dc restricts appliance choice, but an inverter (that converts a lower dc voltage to mains ac voltage) enables the use of mains voltage devices. That, and improvements in appliance efficiency, enable solar to power realistic needs. How to work out solar and battery capacity needed, etc. is covered in Chapter 23.

Promises - promises

Many RVs are intended for the rental-like usage of just coping without mains power for one night (but rarely two). Even then, they may run short of power if a microwave oven is used for more than a few minutes. Their intended usage is of driving several hours each day, stopping for lunch, staying mostly overnight where there is mains power, and an occasional single night away from it.

Most such RVs rely on the associated batteries being alternator charged whilst driving, or by a battery charger and a caravan park's mains power overnight. These systems are increasingly being supplemented by solar, but it is still rare for there to be enough generating or battery capacity to extend to two nights.

Converter systems

For most pre-2000 or so caravans and motorhomes, increasing time on site is remedied by adding or increasing solar and battery capacity, plus the change to the alternator charging system described below.

This is less feasible with post-2000 RVs as some 80% of all locally made (but less so with camper-trailers) have US 'converter' type 12 volt electrical systems that change 230 volts to an unregulated nominally 12 volts dc. In practice the output is usually 13.65 volts on light loads. All lights, pumps and other 12 volt appliances are powered directly (not via a battery) by that unit when a mains 230 volt supply is available.

A rechargeable battery, trickle-charged by a converter, is included but can only power the system for a (typically) single night. The battery is automatically recharged by the RV's alternator whilst driving but unless a dc-dc alternator charger (Chapter 22) is installed such charging can be very slow. The converter's charger also assists if 230 volt power is available, but may take a day or more to fully recharge the associated (typically 100 amp hour) battery. One US converter vendor warns that it may require over three full days. In reality, so-equipped RVs are presumed to have access to mains power most of the time - to the extent that many US RV makers refer to the 12 volt back-up as 'emergency' power.

Solving the issues

It is often suggested to recharge from mains power by replacing the converter with a large high quality multi-stage battery charger, but for most converter-equipped RVs this only partially assists. This is because whilst such converters typically run at 13.65 volts, 12 volt lights and appliances need 12.7-11.8 volts. Some RV makers exploit this by using wiring so thin that 0.75-0.8 volt is dropped between the converter and all running from it.

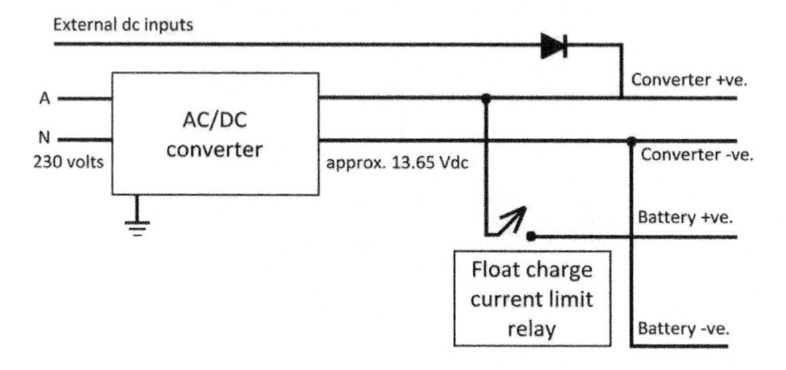

Figure 1.21. Typical ac/dc converter. The unit accepts 12 volt inputs but the series diode drops that by 0.2-0.6 volts. Pic: rvbooks.com.au.

This matters less when the system runs from 230 volts but, on battery operation, appliances may receive only 12.2 volts from a fully charged battery, and a close to unusable 11.6 volts with 50% charge remaining.

A partial solution is to upgrade charging and fridge cabling and replace the existing battery by a lithium (LiFePO4) battery and a charger known to be 100% LiFePO4 compatible. This will charge them to 70%-75%. This will also extend time away from 230 volts because, unlike lead acid batteries, LiFePO4s can be discharged routinely to about 20% remaining. Furthermore, their voltage remains almost constant as they discharge (typically from 13-12.9 volts on RV-type loads).

The only fully satisfactory solution is to replace the converter by a conventional high quality battery charger and install much heavier cable in all but LED lighting circuits (LEDs are less voltage conscious).

In essence, life on generated, stored and then used electrical power is akin to handling your money. Consistently having slightly more coming in than drawn out is fine. The opposite inevitably causes problems.

Generators

Generators are costly to run, and fellow campers in otherwise quiet sites are often bothered by their noise and fumes. Many campsites and national parks now ban them, or have a distant area reserved for their use (often shared by tour groups likely to be even noisier).

Figure 2.21. Honda 2000i quiet generator. Pic: Honda.

Honda and Yamaha have quiet units, but even these disturb otherwise silent bush nights. Those of 2 kW and above will run air-conditioning, but are not intended for continuous use. Most provide mains power and 12 volts dc.

The 12 volt output of most such generators is unregulated and usually limited to 8 amps. Even if labelled 'charger' the voltage output is too low to fully charge. Instead, use a multi-stage mains charger run from the generator's 230 volt output.

Onan has reliable and quiet generators that may be built-in. They are costly and mostly seen in the larger motorhomes and fifth-wheelers.

Wind power

Wind power is only practical for long stays on exposed sites. To be effective, the propeller needs to 35 metres above ground and free of obstructions, a virtual impossibility for RV users. Few small such units produce useful energy below sustained winds of 20 km/h. Most need from 25 km/h upward and that's rare in Australia. They work better on boats and the larger units are fine for property systems, but the necessarily smaller units are of little use for RVs.

Fuel cells

Fuel cells generate electricity from hydrogen currently derived from methanol, LPG, etc, and eventually directly from hydrogen itself. This technology is growing rapidly in the USA with local storage and distribution of hydrogen becoming increasingly available - particularly in California.

Rather than burning fuel to generate energy, operation is primarily electrochemical. It is silent and virtually pollution-free. Fuel cells can to some extent be seen as 24/7 battery chargers. They directly run loads up to their maximum output. Adding a small battery (such as a car starter battery) enables them to run short-term heavy loads - e.g. tyre air compressors and microwave ovens.

EFOY fuel cells, now made in various sizes, are readily available, but require special sealed expendable canisters of ethanol fuel that are difficult to transport. The fuel cells, together with the canisters, unfortunately almost doubled in price by 2016, to a level that is really too high for the general RV market. A major change in distribution however resulted in prices dropping appreciably.

The seemingly promising LP gas-powered unit from Truma was offered for sale in 2012 but, its price (of some 10,000 Euros) proved far too high for all but limited acceptance. It was withdrawn from sale in 2014.

Fuel cells are nevertheless a development to watch. They will inevitably become affordable for RV use as such usage is much the same as the huge markets in third world countries. That usage (for RVs) may well become obligatory for fossil-fuelled cars.

Chapter 22

Batteries & charging

Lead acid: these are the original rechargeable type of batteries and have been in use for 150 years. They were often called 'wet' batteries as they needed topping up with water every month or two. Those used for engine starting were mostly replaced by sealed versions (in the 1990s), as shortly after were all but a few deep-cycle equivalents used mainly in large property systems.

Lead acid starter batteries supply very high current in short bursts, but cannot withstand ongoing deep discharging. They are the weight lifters of the conventional battery world.

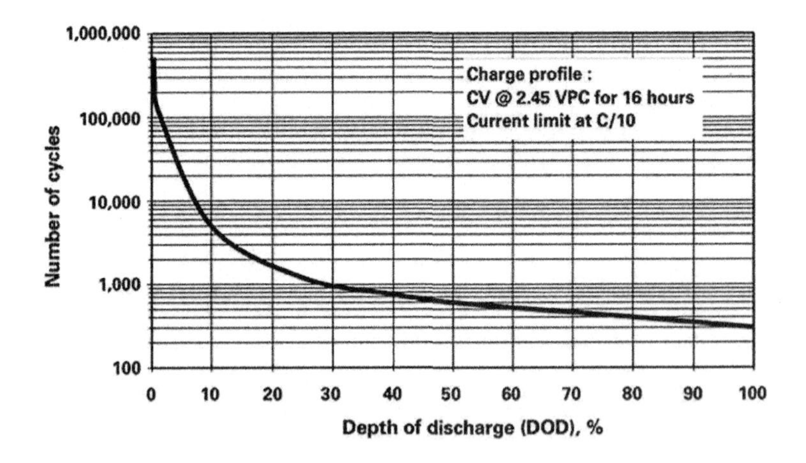

Figure 1.22 Typical number of cycles for a deep cycle battery at varying depth of discharge.

Deep-cycle lead acid batteries withstand deeper discharging than do starter batteries but their life is reduced if often discharged below 50%. They are the marathon runners of the battery world and have long been used as 'house batteries' in RVs.

AGM batteries: an abbreviation of 'Absorbed Glass Mat' AGMs are a type of lead acid battery originally made for military use. The electrolyte is held within a woven mat of thin glass fibres. This ensures enough surface area to contain electrolyte on the cells for their lifetime. These batteries are electrically and physically rugged. They are used mainly in deep-cycle applications and hold a charge far longer than do lead acid deep-cycle batteries.

AGM batteries began to be used in RVs in the mid-1990s. They rapidly gained acceptance, due in part to low initial pricing, but prices escalated from 2007 onward. They can be routinely discharged to 30%-40% (remaining) with less shortening of life. A 125 Ah AGM battery performs much as a 150 Ah conventional deep-cycle battery, but costs about 50% more.

Gel-cell batteries: these preceded AGM batteries and are still used, particularly in large solar systems. They still have a following, but AGMs offer similar benefits and are more rugged.

Lithium batteries: the LiFePO4 version of these batteries is rapidly gaining acceptance in RVs. Most are multi-purpose. All can be used as deep-cycle units and withstand routine discharging to 10%-20%. They also accept and deliver current at massively high levels if required. Charging necessitates a battery management system - including to control individual cell voltages to ensure they are equal.

Depending on load, a LiFePO4's delivered voltage remains almost constant as it is discharged. In RV use it is likely to remain within 13.1-13 volts on light loads and 13-12.9 volts if a microwave oven is used frequently. Some LiFePO4 makers include a full battery management system and suggest a simple two stage charger is then all that is required.

Many battery charger makers are now including a LiFePO4 charging mode in their products. There are also specialised dc-dc alternator charge controllers that include LiFePO4 in their programming.

A major appeal of LiFePO4 batteries to RV users is that such batteries are about one-third the size and weight of others of similar capacity.

Battery life

For lead acid, AGM and gel-cells, their intended life-spans requires them to be routinely fully charged, and not routinely deeply discharged.

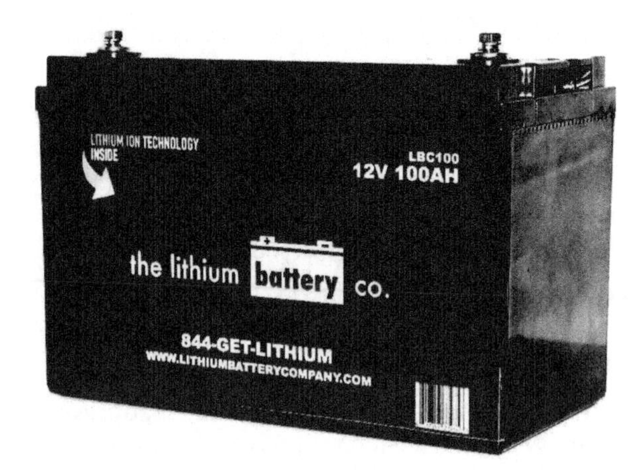

Figure 2.22. Twelve volt 100 amp hour LiFePO4 battery.

Makers of such batteries essentially sell amp hours: you can use a lot quickly or a few slowly. They suggest discharging to 50%. Many users, however, discharge to 20%-30% remaining, drastically shortening battery life.

LiFePO4 vendors now claim about 2000 cycles of typical usage: this is lower than initially, but user reports indicate it to be realistic.

Battery choice

For typical RV on-road use, sealed deep-cycle lead acid batteries are still a good choice, but must be charged correctly (see below) and only rarely discharged below 50%.

For off-road use, AGM batteries are a good choice as they withstand vibration to a greater extent. They are however heavy: a typical 100 amp hour AGM weighs 33 kg. This is not a problem for a bank of them in a coach conversion, but very much so for caravans and the smaller motorhomes that have limited weight allowances.

LiFePO4 may eventually challenge the sales of more conventional batteries, but right now (late 2109), non-technically minded owners are still advised to await direct 'drop in' replacements. Or buy only from a supplier that truly understands lithium technology and seek advice re their charging.

Battery charging

Batteries are charged by feeding them from a supply that has a voltage higher than that of the charging battery. The greater the voltage difference the faster and deeper the battery charges (but has to be controlled as overcharging damages all batteries).

Basic alternator charging uses a constant voltage. The charging battery's voltage rises towards that voltage. As it does so, the difference between the charge voltage and the battery voltage tapers off. Charging current begins to fall at 50% or so of full charge. By 75% it is down to a trickle. This precludes batteries overcharging in vehicles (such as taxis) that operate more or less non-stop, but limits charging lead acid batteries much beyond 75% in normal usage. It does not prejudice starting because the system is designed to cope. Further, due to their construction, starter batteries have far greater charge acceptance. That charge required for restarting is replaced within two to three minutes of the engine running.

Electricity resists moving - it heats conductors it flows through them, losing voltage in so doing. This is not an issue for a battery that is within half a metre or so of the alternator. It is, however, a major problem with camper-trailers, conventional and fifth-wheel caravans, and long motorhomes. With these, alternator to battery conductor runs may total 10 to 20 metres (i.e. one positive and one negative lead).

A battery distanced from the alternator may have voltage barely adequate to charge at all. Heavy cable reduces voltage loss but that really required is starter-motor cable size (that costs $20 or more per metre), or the dc-dc charging described below.

Dc-dc alternator charging

Dc-dc alternator charging largely overcomes the above voltage drop and other problems. It accepts whatever voltage is available (usually from 9 to 18 volts) and, by 'juggling' amps and volts, converts it to that optimally required for the type and capacity of battery charged.

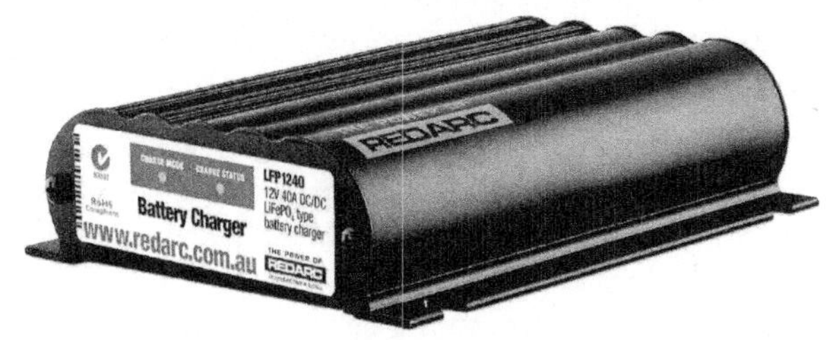

Figure 3.22. Redarc's LiFePO4 (40 amp) dc-dc alternator charger. Pic: Redarc.

Also, rather than charging at constant voltage, dc-dc alternator charging constantly increases charging voltage as battery voltage rises. It charges at constant current, typically halving charging time and ensuring close to 100% full charge.

As some energy is still lost as heat in interconnecting cable, this may limit the maximum charge current otherwise available from the alternator. It is thus advisable to upgrade charging-circuit cabling.

Dc-dc charger units need to be close to the battery/s they charge. It is feasible to use one dc-dc charger in the tow vehicle (if an auxiliary battery is located there), plus a second in the trailer for its associated battery/s. The available alternator output current will be shared, but whatever is available will be at full charging voltage.

A typical 12 volt alternator and dc-dc alternator charger will usually supply 30-40 amps. Allowing for losses it will charge a close to flat 100 amp hour (conventional) in four hours driving. Dc-dc systems are costly but work so well it is pointless to use anything else for RV vehicles and systems of any age.

Battery capacity & losses

Many electrical RV problems are a direct result of not realising that (as with money) unless excess energy is available it is not there to store (let alone then use). Despite this, many people experiencing ongoing 'flat battery' issues attempt to fix this by adding more battery capacity. Doing so is counter-productive. Apart from LiFePO4's 5%-10%, most other batteries lose 15%-20% when charging/discharging. A battery that supplies (say) 100 amp hours requires an average of 115-120 amp hours at 13.6 volts to charge - but delivers 100 amp hours at an average of 12.5 volts.

Battery configurations

Twelve volt batteries larger than 100 Ah are heavy and unwieldy so many owners prefer to have two or more connected in parallel. The batteries must each be of 12 volts but can be of different capacities. Each provides energy according to its capacity and each charges according to its needs. It is also feasible to series-connect two smaller capacity 6 volt batteries to form a higher capacity 12 volt battery.

What does not work is to parallel-connect deep-cycle and starter batteries (or AGM batteries) unless separated by a voltage sensing relay that ensures all initial charge (for two to three minutes) being taken by the starter or AGM battery. It is also not feasible to parallel charge any other type of battery and a LiFePO4 battery because the latter has *very* different charging requirements.

Engine computer issues

Dc-dc charging is perceived by the alternator and engine computer system as a harmless load (much as a pair of spotlights or a hi-fi system). It also overcomes the problem that anything electrical that can be seen as affecting the alternator tends to be blamed if the vehicle has a computer-related problem.

Future alternator charging

Historically, alternator voltage varied from 13.8 volts to 15 volts. This changed around 2000, to alternators that varied voltage to suit the load or were temperature regulated. It is expected that the upper limit may soon become 12.7 volts - some already are.

It is currently unclear how much longer alternator power will be available for RV use. Generating more energy than the vehicle requires for its primary needs necessitates burning more fuel and hence increases emissions. It is likely that its use for auxiliary purposes by RV owners may eventually be banned. Fuel cells may then be used instead.

Mains battery chargers

Low-priced mains battery chargers are a compromise between cost and performance. They are fine for charging sufficiently to start an engine, or slowly charging house batteries. Most charge quickly to 55% or 60%, then taper off. A few, however, just 'go for it'. Sooner or later a cheap one will wreck a battery.

Mains-charging costly batteries necessitates a high quality multi-stage charger. A good 15 amp unit will outperform low price 25 amp chargers and will safeguard your batteries whilst charging. Beware of eBay look-alikes, good three-stage 15 amp chargers cost $200 or so upwards.

Knowing the state of charge

Conventional deep-cycle batteries are snail-like in their response to charging and discharging, particularly when they are cold. Instant voltage readings invariably result in assumptions so flawed that perfectly sound batteries are thrown away, or those close to useless retained.

Figure 4.22. The Victron energy monitor unit is compact and easy to set up and use. Pic: Victron.

A still well-charged battery that has just run a microwave for a few minutes is likely to show only 11.6 volts. That 11.6 volts, however, is also approximately 80% discharge on a well-rested deep-cycle battery. Conversely, a near to useless battery may show 13-14 volts within minutes of being placed on charge, but will drop to 11 volts or less after a few minutes on load.

With LiFePO4 batteries, voltage readings have virtually no meaning. A close to 100% charged LiFePO4 may indicate 13 volts or higher on light load - and still much the same when 30% charged.

The only fully practical way of knowing remaining charge for any rechargeable battery works much as we keep track of money. Count what comes in, count what goes out, deduct one from the other (and the bank's ransom for their use of it meanwhile). The result is what we have left.

Energy monitoring works like that. It is often included in good quality solar regulators but can be added to existing systems - with or without solar. The unit shown (above left) is a good example.

The need for energy monitoring is not obvious to first time owners. It is marginally less necessary if you spend most time where there is mains power but close to essential if you intend to free camp.

Chapter 23

Solar energy basics

It is light, not heat, that solar modules convert into electricity. The amount of solar energy converted depends on the modules' efficiency, the intensity of that light and for how long each day. Most solar modules dislike heat. They lose about 4% for every 10°C above 25°C. They thus work best in cold places under a bright sun.

Figure 1.23. The author's once-owned rig had two separate solar systems, one in the Nissan 4.2 TD Patrol, another in the TVan - seen here crossing Mitchell Falls (Kimberley). Pic: rvbooks.com.au.

On a clear summer midday, the solar energy falling on much of Australia is equivalent to 800-1000 watts per square metre. Depending on their type, commercial solar modules currently turn 14% to 20.5% of that energy into electricity.

Whilst still far from efficient, solar is nevertheless clean and silent and, once installed and paid for, there's free power for a long time.

Storing solar energy

Solar energy is produced only in daylight, but as electrical energy is needed also after sundown and during extended cloud cover, RV (and home) solar systems store energy in batteries for such use.

Until recently, the high weight and bulk of lead acid batteries, plus inefficient charging, limited battery capacity. Now, lithium (LiFePO4) batteries resolve weight, size and charging issues - but adequate solar capacity is still required.

Solar energy enables those free-camping to supply their realistic needs for as long as desired: or on a more modest scale, to stay longer on-site by supplementing battery charge. In caravan parks, generating your own power via solar often enables you to stay in their tent camping areas at a lower rental, and/or save the usual $5 per night for mains power.

Watt's possible

LED lights, TVs (excepting mega-sized), small motor-driven appliances, e.g. blenders, fans etc, present little problem. Nor do efficient and correctly installed compressor-type refrigerators. Laptop computers are fine, but desktop machines used all day are a no-no.

What cannot realistically be run from solar for more than a few minutes is anything that generates a lot of heat. It is still better to use LP gas or diesel for cooking and water heating. Hair dryers are borderline acceptable if used only for a few minutes.

The major energy drawer, in all but a mega-RV, is an electric compressor fridge. Three-way fridges draw far too much power to run on solar for other than quick lunchtime stops. Microwave ovens are fine if used only for a few minutes a day - but are better used only where there's mains power. An often unsuspected energy gobbler is an oldish sleep apnoea machine, particularly if it has a humidifier and/or heater - see Chapter 32.

Golden rule

When thinking of installing solar, work firstly at reducing consumption rather than scaling solar and battery capacity for what you have - or are planning. Use LED lighting, and a post 2014 LED TV.

Consider buying a new high-efficiency fridge. This may cost less than the extra solar and battery capacity needed to run an old one. If the fridge is relatively new, its energy draw is likely to be reducible by installing it as shown in (Figure 7.11) in Chapter 11.

Locating the modules

For RVs, solar's main limitation is the roof space available for solar modules, but is easing as solar efficiency increases. Currently-made solar modules of 200 watts take up only half the space as many a 100 watt module of a decade ago.

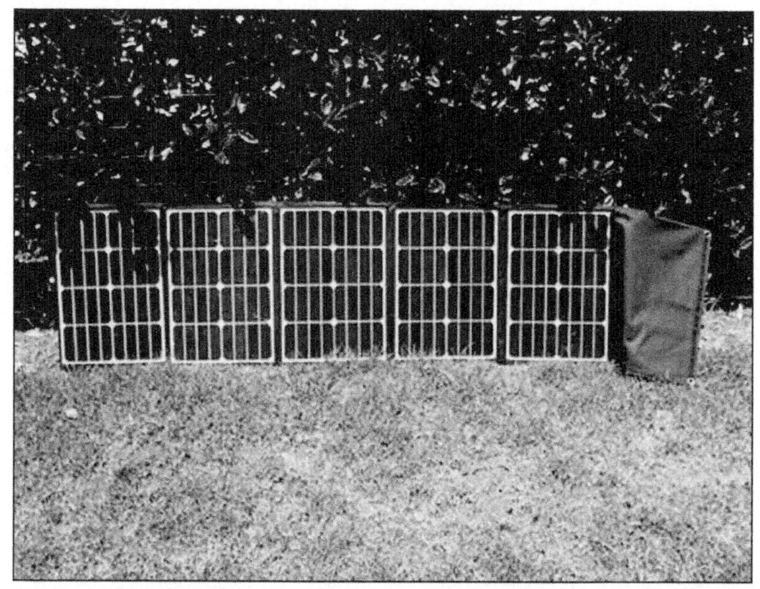

Figure 2.23. Portable 300 watt solar array - fine if people around to keep an eye on it - but otherwise only too attractive to thieves. Pic: source unknown.

Roof mounting solar modules is cheap, simple, effective and virtually thief-proof. It requires the RV to be in the sun, but the required air gap beneath the modules and the roof provides some heat insulation.

Having all or part of the solar array on the tow vehicle (as in Figure 1.23) enables the trailer to be out of the sun. Likewise, a motorhome pulling a trailer can have some or all of the modules on the trailer's roof. A truly effective approach is to have each system fully independent (but inter-connectable) if ever required.

For batteries located in camper-trailers, caravans or fifth-wheel caravans, a dc-dc alternator charger or battery management system also located within such trailers is virtually essential (Chapter 22). It is also strongly recommended for campervans and motorhomes.

Carried loose

Freestanding modules enable the vehicle to be in the shade, but are more readily stolen (Figure 2.23). They work well in isolated places, but even there it is best to have most of the modules permanently located.

Weather effects

Maximum charging often occurs on sunny days that have very light haze or low scattered white cloud, and/or when you are close to an ocean. Such conditions often result in additional sunlight being reflected back up, and then down from underside of the haze or scattered cloud.

Figure 3.23. This bushfire (near Sydney) reduced the output of our solar system to almost zero for three days. Pic: rvbooks.com.au.

Output falls to 30%-40% of normal on cloudy dry days. Heavy rain reduces input yet more. If there's at least some visible light, zero input is rare. It can, however, be totally blocked by smoke from bush fires (Figure 3.23).

Most solar modules also lose much of their output if a shadow blocks part of the module.

Buying solar modules

Solar modules are made in various capacities. The higher the capacity the cheaper they are per watt. The 120-165 watt size modules are a good compromise between cost and ease of installation.

Compare vendor prices. Those of seemingly identical solar modules vary considerably. Buy only totally known long-established branded products.

Do not buy into claims of 'after sales service'. Apart from needing cleaning occasionally (less or not at all in rainy areas), no service is required. Solar modules are reliable and most last well over 20 years.

Chapter 24 outlines the scaling etc. of RV solar systems. For complete solar installation details please see our book *Solar That Really Works* (or *Solar Success* if for a big home system and/or grid connect).

Both books are written in plain English and cover every detail of solar system design, installation and usage.

Chapter 24

Solar specifics

Having reduced energy usage to the minimum, list all items you really need (and their energy draw.) If the draw is unknown use that of the typical appliances shown in Table 1.24. Older units may draw more. Consider updating older appliances as the cost of solar to drive the originals may exceed their update cost.

Most appliances are rated in terms of watts used, but a few are rated in terms of the work they do - not of the energy drawn (in watts) whilst doing so. The latter is *always* higher.

Item	Watts
CD/DVD player	30
Coffee grinder	75
Computer (laptop)	20-30
Computer (desktop)	50-500
Computer printer (ink)	100
Computer printer (laser)	1000
Fans 12 & 230 volts	50
Fridges - see pages 53-54	-
Lights 12 volt halogen (each)	20-35
Lights LED (each)	4-8
Macerator	300
Microwave oven (800 watt)*	1200
Radio	15-20
Sewing machine	75-100
Stereo	50-60
TV (LED) 10-14 inch	20-40
TV (LED) 16-20 inch	40-60
Washing machine (cold cycle)	200
Water pump (12/24 volt)	50-100

Table 1.24.Energy draw of typical lights and appliances.

A microwave oven's wattage rating is of heat produced. The energy drawn in doing so is typically 50% more. A typical '800-watt' unit draws about 1200 watts, or via most inverters, about 1350 watts.

That oven's draw of 337 watt hours (for a typical) 15 minutes is a lot for small systems. The oven may cost only $199, but up to $1000 for the solar, inverter, and battery capacity to drive it. That is a high price for de-thawing an occasional frozen chicken!

To estimate daily usage, multiply the wattage of each unit by the time (in hours or part hours) of its use. For example, a single 5 watt LED draws 5 watts. If three such LEDs are run for one hour, the usage is thus 15 watt hours (often shown as 15 Wh).

Next, from known data, or that shown here, enter that data onto your own version of Table 2.24.

1. In column A, list all lights and appliances.

2. In B, enter the wattage, or total if more than one (such as lights) will be used at the same time.

3. In C, enter hours typically used daily (not rare maxima).

4. Multiply each entry in B by the respective entry in C.

5. Enter the amount totals in D.

6. Total all the entities in D and add 15% (as shown) for charging/recharging losses.

The total shown in 'D' is the minimum amount of energy that needs on average to be generated each day to counterbalance that used.

The additional 15% allows for (lead acid) battery losses incurred when charging/discharging. LiFePO4 batteries incur less. Assume 7.5% to be on the safe side.

Table 2.24 is a typical example of usage. It highlights just how much electrical energy is needed to run an all-electric fridge and also a microwave oven.

Column A	Column B	Column C	Column D
Device	Watts	Hours/day	Watt hours per day
CD/DVD player	30	2.0	60
Coffee grinder	75	0.05	-
Computer (laptop)	30	2.0	60
Computer printer	100	0.2	20
Fan	50	3.0	150
Fridge (electric)	100	10	1000
Lighting (6 LEDs)	35	4.0	140
TV (20 inch LED)	35	4.0	140
Microwave oven)	1200	0.25	300
etc.			
etc.			
Total of Column D	1655		1870
Add 15%			280
Total daily draw			2150

Table 2.24. Make up a table like this for your own usage.

There are no absolutes: an all-electric fridge is convenient to use as indeed is a microwave oven. But if you do decide to have that oven, it makes sense to use it only when mains power is available. Doing so saves the cost of an inverter and the extra battery capacity needed to drive that probably rare high current load.

Solar available

The solar industry (but not academia) denotes the sun's energy using the industry's own derived units of Peak Sun Hours - PSH. Think of sunlight as rainfall and measured as via a solar equivalent of a rain gauge, i.e. by the total amount captured that day. Filling that gauge may take two/three hours much of the day, but on summer days around noon it will fill in an hour across most of Australia. To capture 1 PSH during Hobart's mid-winter, however, requires most of a day.

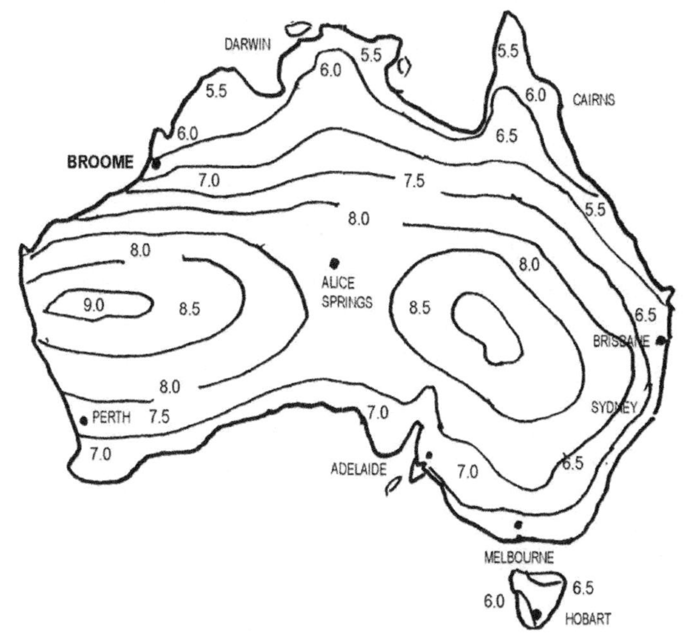

Figure 1.24. This map shows (2015) PSH averages for an Australian summer, but it is safer to assume 10%-20% lower to allow for years of less sun. It is based on NASA data. Map: rvbooks.com.au.

In technical terms, one day's PSH is the equivalent number of hours per day when solar irradiance averages 1000 W/m^2. For example, 6 PSH is the solar energy available over daylight hours that equals the energy that would have been received had the irradiance for six hours been 1000 W/m^2.

If travelling only between October-May (anywhere in Australia) assuming 4-5 PSH provides some allowance for less than average periods.

For winter 3-4 PSH can be assumed for areas north of Brisbane.

Apart from the two mid-winter months in Victoria, there is likely to be 2-3 PSH during the remaining winter months.

What solar modules produce

The solar industry has, in effect, two sets of scales. Standard Operating Conditions (SOC) is for research and marketing. Nominal Operating Cell Temperature (NOCT) is closer to product reality. The substantial difference between that apparently *claimed* (SOC) and that typically *delivered* (NOCT) results in the most probable maximum output being 70%-80% of that seemingly claimed.

A '100 watt' module will produce 70-80 watts along most of Australia's east coast for some part of the day during a typical summer day. If assuming 5 PSH that's 350-400 watt hours/day.

With an electric fridge, one ideally needs 3.5-4 PSH. If space allows, it can be done with fewer PSH by increasing solar capacity, or using a back-up generator, or if affordable, a fuel cell for the two mid-winter months down south.

If roof space for solar is lacking, fit what you can and use a quiet 2 kW Honda or Yamaha generator and a top quality battery charger 40 amp charger to charge the battery bank when solar input is lower. This is desirable anyway as one really needs some form of back up if using an electric-only fridge running from solar in areas likely to cloud over, to rain or (worse) have virtually zero solar input.

Real savings can be made by using a three-way fridge running on LP gas whilst away from the 230 volts grid. This close to halves energy draw. Such fridges cost more initially and there are ongoing costs for gas, but as you no longer need a generator, nor such large battery capacity, the saving is still appreciable.

Solar regulators

Solar regulators control the output from solar modules, ensuring that batteries charge quickly and deeply, but are never overcharged. This enables them to be left on permanent 'float' charge if, for example, sealed lead acid batteries are installed. LiFePO4 batteries require a specific charging routine and it is essential to have a solar regulator that includes that.

There are two main types of solar regulators. The simplest and cheapest, Pulse Width Modulation (PWM) regulators cannot capture and process more than about 71% of the solar input. But as solar capacity is now cheap, for smaller systems it makes sense to accept such loss and add about 10% more solar capacity to compensate.

The second type (and preferable) solar regulator has MPPT (Multiple Power Point Tracking). Whilst often promoted as *increasing* solar output, MPPT recovers, 10%-15% of that otherwise lost. The MPPT function is built into almost all high quality solar regulators. It makes sense to use a high quality MPPT unit with any solar system of above 200-300 watts. Be aware, however, that the MPPT function is often claimed for ultra-cheap eBay specials - but is not included.

Good MPPT solar regulators cost $250 upwards and include monitoring functions. Don't skimp here. It makes no sense to spend thousands of dollars on solar modules and batteries and then throttle them for the sake of an extra hundred dollars.

Most solar regulators need adjusting for battery type, voltage, capacity and time of day. This is not difficult to do - once the manual has been read a few times.

Solar regulator operation is included in dc-dc alternator charging systems - such as those from Cetek, Projecta, Redarc, etc.

Battery capacity for solar

It is pointless to have more battery capacity than you can comfortably charge. The greater the battery capacity, the greater the charging losses. That, for most batteries is 20% or so. That for LiFePO4 is a claimed and probable 5%-7.5%.

Most existing RV systems have too little solar capacity and too much battery capacity. A well balanced system will bring the battery bank to 100% charge (for lead acid batteries a typically indicated on-charge 14.4 volts and a well rested off-load 12.8 volts) by midday most of the year around.

Excess solar capacity

It is good have some excess solar capacity. Much of the time it will not be needed (and the solar regulator blocks that unneeded to ensure batteries will not overcharge) but that otherwise excess is invaluable during long periods of partially overcast sun.

For both battery and solar capacity Table 1.24 is a good guide. Do not exceed that maximum battery capacity as, if done, they may not full charge and their life can be shortened.

Solar capacity (Wh/day)	Optimum battery capacity (Ah)	Maximum battery capacity (Ah)
250	100	125
500	150	200
750	250	300
1000	300	400
1500	450	600
2000	600	800

Table 3.24. Practical battery capacity for solar. Source: rvbooks.com.au.

Solar installation

Solar installation is beyond the scope of this book. It would double its size and increase its cost, yet not be needed by most readers. If intending to do it yourself, my books *Caravan & Motorhome Electrics*, *Solar That Really Works* and *Solar Success* (for the larger home systems) describe how to do it in detail.

Chapter 25

Mains power

Defined in Australian Standards as 'Low voltage', Australia and New Zealand's mains power has been legally defined as 230 volts (+10% -6%) since 2000. It is typically 238 volts in Australia, 230 volts in the EU and 110/220 volts in Canada and USA. In Australia and New Zealand, Low Voltage covers from 50 volts ac and 120 volts dc, and up to 1000 volts ac and 1500 volts dc. From thereon it is called High voltage.

The 12/24 volts that almost everyone calls 'low voltage' is legally known as 'Extra-low voltage'. It defines that below Low voltage. There are variants of Extra-low voltage, but that noted above applies to cabin and RV use.

These terms cause problems for technical writers. If used correctly, readers may be (dangerously) confused. If not used correctly, writers are deluged by complaints from those who know. This section thus uses 'mains voltage' to imply 110-230 volts ac, and 12/24 volts to imply 'Extra-low voltage'.

Mains voltage is dangerous stuff, particularly outdoors. Regulations there are even more stringent. Within Australia, all associated work in electrical installations must legally be done by a licensed electrician or contractor, who is legally liable for its certification (but see below regarding Victorian-made RVs).

Major changes to electrical standards were made in 2000, and again extensively in 2008 and 2018. The main standard is now AS/NZS 3000:2018. That specific to caravans and motorhomes remains as AS/NZS 3001:2008 as Amended in 2012.

The standards apply (with minor variations) to both Australia and New Zealand. They relate to all installed mains voltage wiring, whether the power is sourced from the mains, a generator or an inverter. This applies even if the vehicle has no provision for mains connection. It is, however, legal to run appliances directly from a socket outlet on a generator or inverter intended specifically for that use.

Caravan and motorhome fixed mains wiring differs from domestic practice. For all RVs, circuit breakers, switches and switched power outlets *must* have double-pole switching (i.e. the switches must break the neutral as well as the active lead). This protects users in the event of the incoming power's active and neutral leads becoming transposed. This can happen when people (illegally) make up their own supply cables, or by a (rare) crossed-over connection at a source's supply outlet.

Supply cables

Australian caravan parks have 15 amp socket-outlets with circuit breakers to protect against overload. Residual current devices and circuit breakers disconnect power if specific faults are detected. New Zealand's caravan parks have similar outlets but some also have (still legal) 16 amp and 32 amp outlets.

Current (amps)	Cable area (m^2)	Maximum length (m)
10	1	25
10	1.5	35
10	2.5	60
10	4	100
15	1.5	25
15	2.5	40
15/16	4	65

Table 1.25. Permitted supply cables (for normal loads). It is Table 5.1 of AS/NZS 3001:2008 as Amended 2012.

In both countries, various length supply cables may be used (Table 1.25). Longer cables have heavier conductors. Heavier cable can be optionally used for shorter lengths to provide better mechanical protection. Compliant supply cables are rated such that, at each cable's maximum load, the obligatory circuit breakers will cut the power off within the 0.4 second necessary to realistically save lives. Supply cables must not be interconnected as they may then not be able to cut off the power within that critical 0.4 second.

The 10/15 amp dilemma

A 15 amp plug has a larger earth pin that will not fit into 10 amp socket outlets. This prevents 10 amp cables being overloaded by appliances that exceed 10 amp draw but presents problems where power is required in places that have only 10 amp outlets (particularly in homes). Apart from installing a 15 amp socket outlet (which only solves the 'at home' problem) there are three safe and legal solutions - but all must limit current draw to 10 amps - as people may not legally or safely draw 15 amps via a 10 amp plug.

Figure 1.25. Ten amp RV inlet sockets are now available.

The first solution is to install a 10 amp socket inlet. A 10 amp equivalent is now available that is a direct replacement for the existing 15 amp socket - Figure 1.25.

The RV's existing 15 amp circuit breaker/s and RCD (Residual Current Detector) too must be replaced by 10 amp equivalents. As the draw is now limited to 10 amps, a 10 amp cable can be plugged (physically and legally) into either a 10 amp or a 15 amp socket outlet.

An alternative solution is to scrap the 15 amp inlet socket, change the circuit breaker/s and RCD (as mentioned above) and hard-wire (i.e. direct connect) any cable that complies with Table 1.25. That cable must be mechanically secured at the RV end.

Yet another solution is an adaptor that limits the RVs current draw to 10 amps. The only one currently (2019) legal for outdoor use is the Ampfibian. The unit has a short 15 amp cable that plugs into any 15 amp socket outlet and has an inbuilt 10 amp circuit breaker that limits current flow to that level. Later units also include an RCD (Residual Current Device) - Figure 2.25. A generally similar unit is available from Jaycar, but for indoor use as it is not waterproof, nor electrically rated to permit outdoor use.

Victorian-built RVs

For reasons that are far from clear, RVs built in Victoria are not classified as 'electrical installations'. Electrical work must meet the AS/NZS Standards, but there is no legal requirement for such production work to be done, inspected or even certified by licensed electricians. Certification is likely to be needed if re-registering the RV interstate.

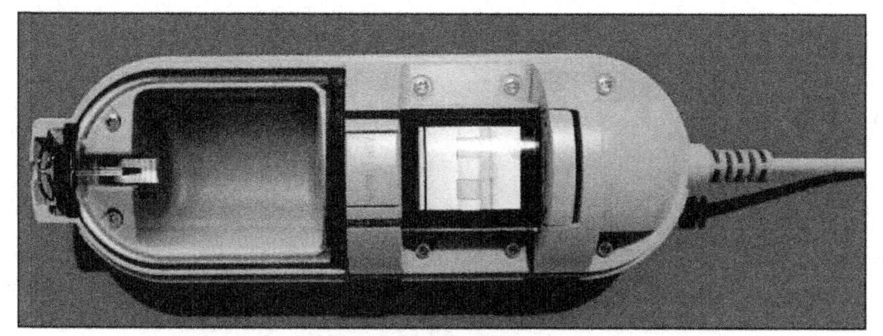

Figure 2.25. The Ampfibian adaptor. Pic: Ampfibian.

New Zealand-built RVs

New Zealand has minor electrical differences. These present no problems for Australian-built RVs in New Zealand, but not necessarily vice-versa. Older NZ-built RVs required earth and neutral to be bonded *within the vehicle*. This link *must* be removed if the unit is used in Australia.

Australian RV practice calls for double-pole circuit breakers and switches throughout (double-pole devices switch or control both active *and* neutral). Single-pole practice is legally acceptable in New Zealand. Such NZ RVs may be used in Australia by their original owners only. They must be brought into compliance before any such RV is *sold* in Australia.

US/Canadian imports

Some imports are claimed to be fully compatible with local standards. In some, however, all that is done is to install a 230-110 volt step-down transformer. As no 110 volt appliance may legally be sold in Australia or New Zealand, such RVs cannot (by definition) be compatible. Furthermore, circuit breakers, switches and socket outlets are replaced (often by single-pole equivalents), rather than by the double-pole versions legally required. Such imports often (illegally) retain 110 volt cabling.

Units primarily at risk are those that have been privately imported - or imported via a local 'facilitator' who arranges the sale etc. The latter may not reveal that the buyer's legal transaction is with the overseas seller. Such buyers may legally use the vehicles in Australia - but cannot legally resell them unless they are made 100% compliant. In some states to do so is a criminal offense.

Chapter 26

Inverters

Figure 1.27 Jaycar M15704 800 watt sine-wave inverter. Lights and appliances are plugged directly into the unit, or via a multi-outlet board. Pic: Jaycar.

Inverters convert 12/24 volt direct current to the 230 volt alternating current required by mains electrical equipment. They are increasingly used in RVs as there is a far much wider range of 230 volt appliances that are more effective, energy efficient and cheaper than their 12 volt equivalents.

Regardless of type, inverters that have socket outlets on the unit (Figure 1.27) must not be connected to the RV's fixed 230 volt electrical wiring. They may only be used by plugging an appliance directly into the outlet socket/s. If in doubt check with the inverter manufacturer.

Sine-wave or modified square wave

A sine-wave inverter's output is identical to the mains supply. The cheaper 'modified square wave' inverters produce 'dirty electricity' and may not run some electrical equipment. Laser printers may be damaged or wrecked by them. 'Simulated' or 'modified sine-wave' units are *not* sine-wave inverters'.

The 12/24 volt input of cheap inverters may not be electrically isolated from their 230 volt output. This can be dangerous in some circumstances. Do not take sales staff advice re this. Few understand 'electrical isolation' and may bluff in response. Electrical isolation is almost always emphasised in makers' literature. If not so-claimed, an inverter is unlikely to be electrically isolated. Buy a fully isolated pure sine-wave unit of a well-established brand. None is cheap.

There are also two main ways of building inverters: transformer based and switch mode.

Transformer-based inverters

Transformer-based inverters produce up to twice their rated output for some seconds and about 150% for up to 30 minutes or so. This is not an overload - they are designed to do this. You can safely switch on (say) a coffee grinder whilst such an inverter is working at its maximum continuous rating. If seriously overloaded, that inverter shuts down whilst it cools.

Transformer-type inverters must be big enough for the job but not hugely more or they will use excessive 'overhead' energy when feeding light loads. The largest load in an RV is likely to be a microwave oven. When powering an '800 watt' oven, that inverter typically draws up to 1450 watts. Allowing for overload capacity, this needs a 1750 watt inverter as there may be lights and a television on at the same time.

Without a microwave oven, a good 500-750 watt such inverter is sufficient for most RV needs. A 750 watt transformer-type inverter will start and run a big power drill or a smallish angle grinder.

Pre-2000 transformer-based inverters were typically 75% efficient. If you have one, replace it as the best current units are now 95% efficient across much of their range - you will spend more on generating capacity and replacing batteries than that new inverter.

These inverters are ultra-reliable but bulky and heavy. They are thus fine for cabins and the larger RVs, but their size (and particularly weight) may be an issue in the smaller RVs.

Switch-mode inverters

The later-developed switch-mode inverters are smaller, lighter and much cheaper per watt. Their downside (in some uses) is that many have no overload capacity (or if so for only for a second or two). For many, their continuously rated output is (as with some small generators) only 80% of that apparently claimed. With some eBay specials, it can be as little as 40% if run for more than a few minutes.

Switch-mode units are fine for appliances such as LEDs, TVs, radios, etc. that use much the same energy to start as they do to run. Big power tools, many air conditioners and appliances that have heating elements or electric motors draw several times their running current for a second or two whilst starting and the inverter must be large enough to supply that current.

In some cases it is cheaper to buy a transformer-type inverter of known high overload capacity. Buy only known brands, or from a well-established vendor.

Chapter 27

Preparing for the trip

Poorly maintained vehicles are a repairer's nightmare. Any one or more long-neglected components can trigger a breakdown, but countless potential causes of breakdown remain. A repairer may clean the gunk out of a blocked fuel line and advise that the whole fuel system needs cleaning and filters replaced. This costs $200 or more, and is often rejected. Yet, when that dirt again blocks the system, the original repairer is likely to be blamed.

Figure 1.27. Carrying spare filters and other routinely replaced components can save days of waiting for parts in remote areas. Pic: Nissan.

Components occasionally and prematurely fail, but most breakdowns are due to the failure of belts, hoses, filters and shock absorber rubbers, etc. that have long exceeded their service life. Regular preventative maintenance and the replacement of worn components is essential.

Fuel

Water, sand, rust or paint flakes find their way into your vehicle's or a fuel vendor's tank. The original filter traps most of this, but it is better to have a second filter installed between the first filter and the engine, (some vehicles have dual filters as standard).

Check first with a workshop specialising in your vehicle. This is because some systems use the fuel to lubricate and cool the fuel pump and may not tolerate additional restriction. Good fuel filters costs around $150. They are a sound investment.

Water in diesel fuel turns into high-pressure steam when injected into the cylinders. This has exploded injectors. To discourage this, diesel fuel filters have water traps. Drain them twice yearly.

Avoid buying unknown-brand fuel - particularly diesel. It may contain solvents, crude oil, etc. that vendors buy free of tax. Such mixtures may cause the engine to overheat and/or foul the injectors.

Dirty or contaminated bulk storage tanks are also occasionally encountered. Where possible, buy from major-brand service stations in large towns, or from truck stops.

Have diesel fuel injectors cleaned and adjusted every 100,000 km, or if the engine blows black smoke under load at normal engine speeds. (Most diesels blow some smoke if flogged at very low engine rpm.)

If the vehicle is off the road for more than three months or more add an anti-bacteriological diesel additive It is available from some truck stops and 4WD parts stores.

Air filters

Engine air filters must clean indrawn air without restricting its flow. Filter elements need cleaning or changing regularly, even if seemingly still new. As filter elements cost $35 upwards, consider replacing them with cleanable foam equivalents using the specially formulated oil supplied by the makers. The special oil traps the dirt - the foam element holds the oil in place.

Oils ain't necessarily oils!

Dirt build-up is not the only reason for changing oil. Most engine lubricating oils have multi-viscosity - thin when cold to ease starting, and thickening as they become warm, but over time they lose this property. A nominally 20/50 oil eventually reverts to a straight 20 grade and is too 'thin' for some engines.

Diesel engines need oil changes at least as often as the manufacturer recommends. If possible, follow manufacturer's recommendations. If you can't find the exact or an equivalent oil, an oil marked 'API CC' or 'API SC', for diesel and petrol engines respectively, will suffice until the next oil change. This does not apply to two-stroke Detroit Diesels. These engines *must* have the oil specified.

Oils are now available with longer lifetimes. Before using check with the vehicle manufacturer about extending times between oil changes - especially if the vehicle is under warranty.

Servicing the electrics

With older petrol-engined vehicles, clean and re-gap spark plugs and contact breaker points every 5000-7500 km. Set the contact breaker gap using a dwell meter, not feeler gauges.

Figure 2.27. Dwell meter. Pic: UNI-T.

A dwell meter (Figure 2.27) measures the cam angle during which the contact breaker points are closed. In a four-cylinder engine, the angle between each ignition cam lobe is 90° and the dwell is usually just 45° of distributor rotation. In a six-cylinder engine, the lobes are 60° apart and the dwell time is 30° to 35° whilst the engine is running. The required dwell is usually shown in the vehicle's full technical manual.

Dwell measurement provides a truer reading because distributor shaft and bush wear cause major discrepancies between static and dynamic gap settings. If adjusted by measured gap settings this may prevent the spark coil generating enough energy to start a cold engine. It can also cause erratic ignition timing.

To measure dwell, bring the engine normal operating temperature and then shut it off. Connect the dwell meter's positive lead to the positive terminal on the ignition coil, and its negative lead to the negative terminal on the ignition coil. Then run the engine.

Replace plug leads and distributor capacitor before heading off, and carry a spare coil, distributor cap, capacitor and rotor.

Computer-controlled engines

Computer-controlled engine systems are now essential to reduce emissions. This has increased engine reliability but, if problems arise, the engine is likely to need expert attention, special testing equipment and cost depressingly large amounts of money. Problems are mostly due to the failure of sensors that provide the computer with engine information. The computer-units themselves rarely fail but they are often wrongly blamed and replaced regardless of their high cost.

Vehicles with computer-controlled engines need to be maintained by authorised agents. For outback breakdowns it is best to truck the vehicle to the closest such facility rather than having it messed around by non-experts. Many outback dwellers and travellers thus prefer basic non-computer diesel-engined vehicles such as the pre-2005 Toyota HiLux, Troopy, the 4.2 litre TD Nissan Patrol and an OKA. These are ultra-reliable if they have been maintained correctly, or totally overhauled, and outback mechanics know them inside out.

General

Good engine compression is essential. Diesels engines *rely* on it to ignite the fuel. Loss of compression is mostly caused by ingested dirt and fine sand becoming embedded, thus abrading cylinder walls and piston rings. Valves and valve-seat burning problems are mostly historical, but may occur on engines made before the 1980s, or if an engine intended to run on leaded petrol is run for long on unleaded fuel.

Spares

If you have a newish, popular, and well-maintained vehicle, and you do not venture far off main routes, you can get by with basic spares. Fan and alternator belts, radiator hose/s, fuses, headlamp globes, plus tyre changing equipment and basic tools will suffice.

Carrying spare parts will save you time and money if you break down. This also saves time and transport costs during routine services. Most other bits are likely to be available at the next major town, or air-freightable within a day or so. Australia's main coach lines carry freight up to 20 kg. They are often the quickest way to ship spare parts.

Older, less common vehicles, or those driven in barely populated areas need more. The recommended lists of tools and spares (listed in Table 1.28 in the next chapter) may seem like overkill, but even if you don't know how to use the tools or replace bits and pieces, you'll find people who can and often will for you.

Caravans and camper-trailers

The most common failure with camper-trailers and off-road caravans is spring breakage and stub axle breakage, wheel nuts loosening until a wheel falls off, damaged or worn wheel bearings, worn-out shock absorbers and shock absorber rubber bushes.

The above is far more common with units that lack shock absorbers altogether, or have faulty or worn out shock absorbers. Some trailers have springs so rock hard that they do not deflect enough to break, but the rest of the RV suffers instead. That adequate shock absorbers are essential has been known since the late 1800s. Most spring, stub axle and wheel stud breakage issues are attributable to this cause.

Chapter 28

Common problems

This chapter lists the more likely problems and how to avoid them. Table 1.28 recommends the tools and spare parts to carry. Items in italics (in that table) apply to vehicles that will be used in extensively in isolated areas.

Air filters: clean or replace. A partially blocked air filter is unlikely to disable a vehicle. It may, however, lead to power loss with diesels, and also increased fuel consumption with petrol engines. Air filters need replacing frequently if travelling on dirt roads. They are better replaced by cleanable foam equivalents.

Battery: replace if more than three years old. If less, have it load-tested and replaced if necessary. Remove and clean battery connectors. Use an anti-corrosion material when reconnecting. Apart from their age, common causes of failure are undercharging, and shock loading caused by loose battery mountings. Battery clamps that nip onto an edge at the base of the battery may fail on corrugated roads. Add extra clamps if necessary to truly secure the battery.

Battery starter: If travelling in outback areas you cannot rely on another vehicle assisting you to restart. Carry a LiFePO4 battery starter - if it is fully charged it will retain that for several years. Most will start a big 4WD diesel engine several times. This is an investment you cannot afford not to make.

Brakes: replace brake pads or linings if more than two-thirds worn. Check brake lines for damage. The brake industry advises to change brake fluid yearly.

Clutch: check that the clutch pedal still has adequate remaining adjustment. If in doubt, consult a specialist. Replace clutch plate if so advised.

Drive belts: check for wear and tear, and adjust tension. Replace if necessary, or if they are more than three years old. Although the battery will keep the engine going for a time, a broken alternator belt will eventually bring you to a halt. Always carry a spare.

Drive shafts and universal joints: using a torque wrench, check nuts/bolts for tightness Lubricating propeller shaft splines is *essential.* If they seize, damaging thrust loads are thrown onto the differential, drive shaft, and gearbox bearings, and the engine and gearbox mountings.

Engine mounts: replace the rubber mounting blocks if damaged, or are over seven years old.

Exhaust pipe/silencer: check exhaust pipe, silencer and associated mounting brackets and rubbers. Replace them if remotely suspect. Ideally, have this done by specialists - it is a tricky and filthy job.

Fuel filter/s: clean or replace. Refer to previous comment before adding an additional filter.

Ingested dirt/water: an engine dislikes ingesting hot air, dust and all liquids except fuel. Many outback travellers add snorkels that take in cleaner, cooler air at roof level. Snorkels are tricky to make and install - go to one of the many 4WD equipment suppliers and have the job done professionally.

Oil temperature: when travelling in the hotter parts of Australia an oil temperature gauge is almost as valuable as a water temperature gauge. The maximum safe temperature for engine oil is about 110°C. Temperatures above this may permanently change its viscosity, possibly leading to bearing failure.

Radiator: check for leaks, also check radiator mountings and replace or repair if needed. Have the radiator reverse flushed if it is more than five years old.

Radiator hoses: hoses in good condition feel springy. Replace any that feel rigid or floppy, and routinely if over five years old. Tighten hose clamps. If hoses are replaced, check tightness again after 100 km.

Shock absorbers: have the shock absorbers checked by a shock absorber testing centre, and replace if necessary. In most cases this can be done whilst the shock absorbers are still on the vehicle. Also replace shock absorber bushes if they are more than two years old, or every 25,000 km if the vehicle or trailer is used extensively off-road. If a trailer has no shock absorbers do not risk going off-road: the chance of major problems is high.

Springs: potential problems are hard to spot. A certain warning is if the vehicle sits close to its bump stops. If in any doubt, have the vehicle checked by a spring specialist. Rectification may require springs to be removed from the vehicle to check and, if needed, to rectify their free-state curvature.

Tyres: replace tyres regardless of wear if they are more than seven years old (see also Chapter 12. Do not use tyres more than seven years old if travelling extensively in outback or isolated areas.

U-bolts: clean U-bolts and nut threads. Re-tension the nuts, using a torque wrench. Re-check after 100 km and 1000 km, then check at least every 2500 km (some outback tour operators check every day). Ditto for trailers. Consult a specialist regarding upgrading to heavier duty replacements.

Wheel bearings: check for correct adjustment, re-pack with wheel bearing grease.

Tools (general)
Allen keys
- metric & imperial
Funnel
Hammers
- light, medium, heavy
Mirror
- small
Pliers
- regular
Screwdrivers
- regular, Philips
Spanners
- adjustable
-small, medium, large
- oil filter wrench
- open-ended
- ring
- sockets
- spark plug spanner
- sump plug
Tweezers
Vice grips
Work gloves

Add (off-road use)
Centre punch
Cold chisel
Cordless drill
Drill & drill bits
Feeler gauges
Files
- flat, half round, rat-tail
Funnel (plus strainer)
Grease gun
Hack saw & blades
Pliers
- circlip

Spares (electrical)
Fuses
Globes
- headlights
- interior
- reversing
- side lights
- stop
- tail lights

Add (off-road use)
Dist. cap, rotor & points
(or complete distributor)
Ignition coil
Spark plugs & lead

Wheels/ Tyres
Tyre pump or compressor
Valves & valve stem tool

Add (off-road use)
Bead breaker
Hi-lift jack
Inner tubes
Second spare wheel
Tyre repair kit
Tyre levers (3)
Wheel nuts/studs (ditto for trailer)

General
Araldite
Cleaning rag
Jumper leads *
Oil
- engine, gearbox
- differential/s
Radiator sealant
Rope

- round nosed
Punches (various)

Tools (electrical)
Crimping tool
Multimeter
Polarity tester

Add (off-road use)
Soldering iron
Timing light

Spares (general)
Air filter/s
Brake linings, disc pads
Fan belt/alternator belt
Fuel filter
Hose - radiator
Keys (secreted under vehicle)
Oil filter
Wheel jack
Wheel wrench (strong)
Tyre pressure gauge

Add (off-road use)
U-bolts (also for the trailer)
Shock absorber bushes (as above)
Fuel pump (or diaphragm)
Fuel cap
Hose (general)
- brake bleeding
Hose clips
Wheel bearings for trailer

WD-40 (or similar)

Add (off-road use)
Brake fluid
Emery paper
Epoxy (two part)
Exhaust putty/tape
Fencing wire (8-gauge soft)
Gaffer tape
Gasket material & cement
Grease
Power steering fluid
Radiator leak fluid
Screws, nuts, etc.
Soft wire
Sealing tape (water resistant)
Wooden blocks (assorted).

Table 1.28. Tools and spare parts to carry. Items in italics apply to vehicles that will be used in extensively in isolated areas. It does not matter if you do not know how to fit them - or even use the tools. Someone inevitably comes along who does. The vital thing is to have the tools and bits available.

Chapter 29

Outback travelling

In Australia, fuel is available within 400 km intervals on most outback routes except in northern WA where there are a few 500–600 km gaps. A range of 700 km gives an adequate margin for the Birdsville, Oodnadatta, and Strzelecki tracks. Diesel is available from fuel stations, (some) homesteads and Aboriginal communities. Opal (petrol) is also available from most. For the Tanami Road (Halls Creek to Alice Springs) fuel is again available at the Yuendumu Aboriginal community (just to the east of Rabbit Flat), and also at Tilmouth Roadhouse, about 220 km west of Alice Springs. There is a gap of about 500 km on the Plenty Highway as fuel is no longer available at Tobermory Station. For the Canning Stock route, seek up to date advice from tour operators (this track is definitely not feasible for any caravan and not that good an idea for heavy camper-trailers).

Diesel usage on dirt tends to be 10%-15% more than on bitumen roads. With petrol engines, however, fuel usage may double in soft going.

Fuel storage

Jerry cans are convenient but stowing them can be a problem. It is unsafe to carry fuel inside the vehicle, or have its weight on a roof rack two or three metres up. Consider having two tanks, each with its own filter/s. Include a changeover valve enabling the existing fuel gauge and pipes to be used for either.

Specialised servicing

Ensure your vehicle is mechanically sound before setting off. Advice on previous pages applies, but unless you really know what to look for, have your vehicle checked by a specialised 4WD service centre. Its staff is more likely to understand your needs than staff from manufacturers' city service stations.

Bull bars

Bull bars may help to protect a vehicle and its occupants in the event of hitting animals, and ease branches aside on narrow tracks. They are unlikely, however, to protect occupants in a front-end crash because (in older vehicles particularly) they reduce the ability of the front end to crumple progressively. In such cases, the bull bar may protect the vehicle but increase the forces acting on its occupants.

A further concern is that a poorly designed bull bar may affect air bag triggering. You legally may not fit a non-approved bull bar to an air-bag equipped vehicle.

If buying, do so only from companies (such as ARB) specialising in this area as they must now be designed to minimise the severity of collisions with pedestrians.

Driving off-road

Until you are familiar with your rig's capabilities and limitations, and that will need a full year or so to gain, do not travel on any soft or slippery surface in lower than third gear (low range) or second gear (high range) or in mud. Reserve 4WD for getting out of trouble - not further into it.

Figure 1.29. Not bogged - simply reducing tyres pressures on part of the Canning Stock Route. Pic: Author – rvbooks.com.au.

It is hard enough if your four-wheel-drive gets bogged, but if that happens whilst towing a caravan you will need to debog the tow vehicle and caravan separately.

Complete a government accredited off-road driving course taught by instructors experienced also with 4WD trucks (some teach bush fire brigade drivers).

Some 4WD clubs run off-road driving courses, but not all have the different skills needed with big units, or caravans.

Driving in sand

Reduce tyre pressure (including a trailer's) to 170 kPa (25 psi). If towing, engage four-wheel-drive and lock the front hubs (if you have lockable hubs). Until you gain experience, limit speed to about 30 km/h.

If the sand softens, you will be audibly aware the engine is working harder. If, especially when towing, the engine starts to labour in third gear low-range (until you are more experienced, or accompanied by someone who is), stop there and then. Walk the proposed route to check for firmness, or make a very wide turn (again, having checked that surface also) and go back the way you came.

After you gain experience, you can drive in second gear (low range) but, if the engine begins to labour, you are generally very close to becoming bogged. If the engine stalls or lacks power to go any further in second gear low range, do not even think of using first gear unless you are truly experienced.

If you do get to that second gear stalling point, stop and reduce all tyre pressures to about 18 psi (125 kPa) then dig away sand blocking the front tyres, and between the front and rear tyres - and also any close to the underside of the vehicle. Then seek a way of making a very wide radius turn. Walk that route to check the surface, very gently restart and remain (in first gear low range) until you are back on the track. Then return the way you came.

If it is not possible to make that turn, check if there is a drivable path around the soft sand ahead. Unless that seems like the end of the soft stuff, make that turn and go back.

If you drive in sand as advised here, you are unlikely to get bogged. If soft sand does drag to a halt, stop the engine *immediately*. Spinning the wheels (novices always do) even for a quarter turn will inevitably bog you deeper.

For a solo vehicle dig a clear flat channel for the tyres some metres to the rear and between the front and rear tyres. Lower all tyre pressures to 110 kPa (16 psi). Ensure (manual) front hubs are locked and four-wheel-drive really *is* engaged and have the front wheels pointing absolutely straight ahead. Then attempt to gently reverse out, but stop the instant a wheel even slips slightly. If necessary use the vehicle floor mats to aid flotation.

If towing a trailer, uncouple it and cautiously find a path to take the tow vehicle to the trailer's rear. Do whatever you can to rotate the trailer such that it can be recoupled - and return the way you came. Another technique is to have a second tow hitch receiver at the tow vehicle's front - enabling it be dragged out by the tow vehicle in reverse.

In soft going, drive in as straight a line as possible. If you absolutely *have* to turn, do so at the widest possible radius and at walking speed. Until the tyres are reinflated, avoid tight turns as there is a risk of rolling a tyre off its rim. Once clear, reinflate tyres, but do not exceed 20 km/h until you do so.

There is a good case for lowering tyre pressures by about 20% and engaging four-wheel-drive in all soft going. It reduces tyre wear, eases the load on the transmission, and (by reducing drag), eases fuel usage.

Do take reducing tyre pressure seriously. An overseas tourist died when her campervan became bogged about 20 km off the Oodnadatta track. When recovering the vehicle, police simply dropped the tyre pressures - and drove straight out.

Beach driving

The best advice is don't - particularly if pulling a trailer. If you really must drive on a beach, stay mostly high up the beach. To turn, drive down toward the sea and gradually edge back up in a wide and gentle curve.

If caught on an incoming tide, also use everything possible under the tyres. This includes door mats and carpets, boot linings, mattresses, even cupboard doors if they can be torn off. Be prepared to sacrifice anything like that. On a rising tide there is no other way of getting out - except for another vehicle and a long snatch strap.

Having seen over ten 4WDs lost on Cable Beach north of Broome (where there is a 10 metre tide) I strongly advise it is better to park the rig and walk.

Debogging in mud

This is hard enough with a single vehicle. If it happens when towing, uncouple the trailer and seek help as there is a real risk of otherwise bogging both. If that happens with a heavy motorhome, only a tractor or a heavy vehicle with a winch or snatch strap will get you out.

Debogging (snatch straps)

Debogging by pulling by another vehicle, or a via a winch, is limited by the power and tyre grip of the tow vehicle or the power of the winch. The debogging force required, however, is readily achieved by using a snatch strap to exploit the inertia (resistance to change) of an already moving vehicle. The strap increases the pulling force by five or more times.

A snatch strap is a length of semi-elastic webbing with a typical breaking strength of 12 or so tonne. It is attached to the rear of the recovery vehicle and to the front of the bogged vehicle - with about two metres left free.

The recovery vehicle is driven away at a brisk walking pace. As it does the elastic strap stores and then releases the kinetic energy of the accelerating vehicle - resulting in a pulling force massively greater than a direct pull. It works much as did Roman catapult 'siege engines' that could hurl half-tonne rocks over 100 metres.

A snatch strap's downside is that it as potentially dangerous as those early catapults. Many non-technical people grossly underestimate the forces involved, and hook the pulling loop over the tow ball. The forces are so high that the tow ball can be torn off and hurled for over half a kilometre: sometimes via the windscreen and rear window of the bogged vehicle. Take this very seriously. Drivers of bogged vehicles are known to have been killed this way.

A snatch strap must only be attached to both vehicles via shackles officially rated at twice that strap's breaking strength. Do not use the non-rated shackles sold by hardware stores, they are of unknown strength.

The shackles must be attached to both vehicles by equally strong engineered eye bolts, or by a method that can be relied upon to be very much stronger than that strap and shackles.

Ideally have a certified engineer design and arrange to fit suitable attaching points for the snatch straps on the front and rear of your tow vehicle.

Using a snatch strap

Dig away any banked up sand, or do what you can to ease movement of the bogged vehicle. Then:

Figure 2.29. How to use a snatch strap safely.
http://i.ytimg.com/vi/QiddIULJVHQ/maxresdefault.jpg

1. Ensure no-one is within 400 metres of the bogged vehicle's rear. Place someone whose *absolute priority* is to keep people at least 10 metres to the side, and children at 50 metres, and carefully watched.

2. Have the two vehicles in a totally straight line, with their front wheels pointing straight ahead, and such that a two metre length of snatch strap is left lying in a loose 'S' shape about a metre or so across.

3. In sand, lower the tyre pressures of the bogged vehicle to about 110 kPa (16 psi).

4. Place a heavy blanket, or one of the devices made for this purpose, folded over the centre of the strap. This dampens some of the huge forces should the snatch strap break.

5. Start both engines. Select the bogged vehicle's first gear with clutch depressed, or 'Drive' if an automatic.

6. Release the hand brake of the bogged vehicle.

7. Drive the recovery vehicle away in a straight line at a brisk walking pace in first gear. The moment the bogged vehicle begins to move, accelerate that vehicle gently.

If this fails, see if anything can be done to free up the bogged vehicle and try again - but do not exceed a brisk walking pace or there is a probability (not just possibility), of breaking the snatch strap or tearing off the attachments. Renew snatch straps after ten uses.

Figure 3.29. This shows how just how effective a snatch strap can be. The author's 5.2 tonne OKA was once bogged in a salt-bed creek. It was pulled out with relative ease, via a snatch strap, by an unladen 2100 kg Toyota Landcruiser ute!. Pic: rvbooks.com.au.

By far the best way to learn to do this with minimum risk is to enrol in one of the accredited off-road training courses run by TAFE and some 4WD clubs. Ensure they *are* accredited.

Chapter 30

Keeping safe

Risks are low whilst travelling in most countries, but precautions are advisable when bush camping. Drive early in the day, and seek a camp site at least two hours before sundown as it is close to impossible to find and assess the safety of a potential site whilst driving in the dark. Check national parks, state forests, etc. even off-route. Also check likewise for a fall-back caravan park. In towns it is usually possible to stay overnight in a supermarket parking lot (but first seek permission from the security guards).

Being hassled is rare, but can happen. It is disconcerting, but is usually bored kids hooning around, and mostly limited to 30-40 km of large towns (particularly on Friday nights). Physical attacks, however are rare.

If travelling off major routes, ideally travel in company with another vehicle. If alone, if you see another RV in a promising site ask if you too could be there overnight as most travellers welcome the emotional and physical comfort.

Avoid camp sites visible from the road. Ensure no light can be seen from outside your vehicle. Do not have a camp fire as the glow on tree tops can be seen kilometres away on a dark night. Never camp underneath eucalypt trees. Heavy branches snap off without warning, even in dead calms.

Always ensure there is a clear exit with no need to reverse. Pack up before going to bed so you can drive off (in a campervan or motorhome) without exiting the vehicle. Ideally have one that has direct access to the driving cab. Lock the entry doors from inside but leave the keys in the inside lock (in case of fire). Likewise the ignition key/s. If threatened, don't venture outside. Drive away and call the police - or pretend to - even if out of range.

Carrying weapons

Not all agree, but my strong advice is never to carry a weapon. If you show a weapon you must be prepared to use it or an attacker will use it against you. In 1960-1962 I drove a big mobile laboratory around Africa: including twice through the Algerian war, the then-Belgian Congo at the time of independence, and uprisings in Rhodesia, South Africa and Kenya. Since then, my wife and I have driven around and across Australia over 12 times via the Centre. We usually bush camp and have never once felt the need for a weapon, nor felt seriously threatened. Our own and many others' experience is that choosing sites at least 45-50 km away from major towns, and in ample time, is sufficient.

Fire

Install two dry powder fire extinguishers. Locate one close by the exit, one in the kitchen area, and with pressure gauges readily visible. Have an immediately accessible fire blanket in the kitchen.

Bush fires

Despite Victoria's bush fires in 2009, most of Australia has a reasonably effective system that ensures roads are closed to traffic during risk of fire. This is not necessarily so in isolated areas but most bush fire prone areas have signs that indicate the current risk.

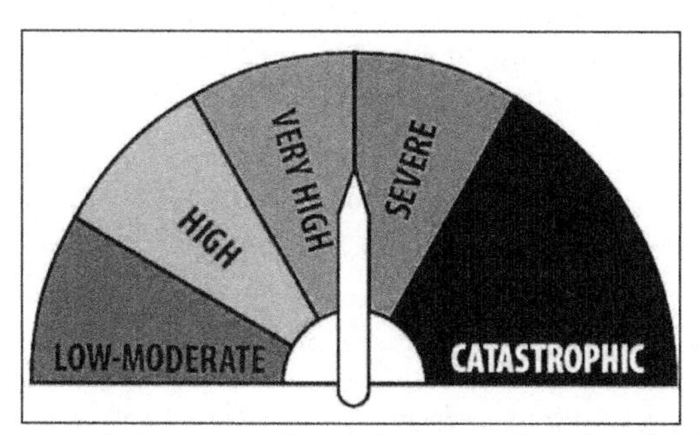

Figure 1.30. A sign such as this, at the entry points of roads in bush fire prone areas indicates the current fire risk.

There are good Internet satellite services that provide maps of previous and existing fires (including spot fires) and are current to within a few hours. The Firenorth service covers the whole of central and northern Australia: firenorth.org.au/nafi3/

Landgate covers all of Australia: firewatch.landgate.wa.gov.au/landgate_Figurfirewatch_public.asp

Insects

Mosquitoes are more than just a nuisance. In Australia, some carry Ross River Fever, Barmah Forest virus, the potentially fatal Murray Valley encephalitis, and also dengue fever (especially in and around Cairns). Peak mosquito time in tropical/semi-tropical areas is between the end of the wet and the beginning of the dry seasons. In other areas it is at any time after extensive rain, particularly when cloudy.

Insect repellents are advisable: some of the more effective contain diethyl-toluamide (DEET) that may cause allergic reactions. There are also 'natural' repellents based on substances such as citronella, tea tree oil, or pyrethrum. All lose effectiveness after a time, so effective screens or nets are necessary whilst sleeping. Also effective are the vapourising devices made by Mortein and others.

Early editions of this book noted that daily 100 mg tablets of B1 were claimed to reduce the incidence of mosquito bites. Many sellers of such products still claim this but subsequent research suggests that this only *masks* the swelling and irritation of such bites. This is a possibly dangerous situation as it may mask the presence of disease-carrying mosquitoes.

Sandflies (midges)

These can cause allergic reactions out of all proportion to their size. The best defence is to avoid areas known to be bad. Local chemists will advise. A good palliative is antihistamine tablets. Start a day or two before entering sandfly areas. Use antihistamine cream at night. Carry these tablets anyway as physical reactions can be so strong that medical attention may otherwise be essential.

A surprisingly effective and cheap repellent is 90% Sorbolene, 10% Dettol, plus a drop or two of tea tree oil and a little of any preferred essential oil.

March flies

These are very prevalent in northern parts of Australia at some times of the year. They are vicious, determined creatures with a painful sting. Their only virtue is that they are slow moving and easy to zap.

Marine stingers

These can kill. Avoid swimming in stinger-prone seas from November and March. Stingose or raw vinegar provide some relief from the less venomous. Pour over the affected area/s but do not rub in.

Crocodiles

Crocodiles inhabit many wet areas north of a line from Bundaberg to Port Hedland. Assume there are crocodiles unless you are absolutely certain there are not. Often there are. There are two main types of crocodile and both are misleadingly named.

Figure 2.30. Saltwater crocodile. This creature is as dangerous as it looks. Take this seriously: they really do kill people. Pic: Aquazoo.

Saltwater crocodiles also known as estuarine crocodiles (Figure 2.30) are seriously nasty: their ambition is to eat you. They are found in the sea and in rivers up to several hundred kilometres inland. They exist for months on end in seemingly dried up river beds, and billabongs little larger than a home spa. It is rarely possible to tell if one is there unless it surfaces. If you are within ten metres it can be too late.

Saltwater crocodiles (Figure 2.30) can jump, even vertically, at extraordinary speed and ferocity, to well over their extended length. They are typically three to five metres long, but may grow to over six or more metres. They study routines and generally attack after someone visits the same spot several times - but you cannot assume you are safe the first time.

Common advice is to camp at least 50 metres from where there may be saltwater crocodiles. Aboriginals advise that 100 metres is safer. They suggest lighting a fire as crocodiles are afraid of it. There is a minor risk of such crocodiles on flooded roads, particularly in the Kimberley. There are ongoing sightings during the wet season.

Figure 3.30. Freshwater crocodiles have narrower heads than saltwater crocodiles and rarely exceed two metres. Pic: Aquazoo.

Freshwater crocodiles (Figure 3.30) are generally smaller (typically 1-2 metres) and have much narrower heads. They are relatively harmless and will usually only attack if provoked, but it is never a good idea to get between a mother freshwater crocodile and her babies.

Snakes

Most snakes tend to coexist but regard being trodden on as unfriendly and react accordingly. They usually slither away some seconds after they sense your presence: reacting to changes in light, shade, movements and vibration.

In snake-prone areas walk slowly and heavily. If concerned, pause every few metres to give them ample time to respond to the vibration. Leave about 10 metres between the leader and whoever follows as snakes often react a few seconds after the first person has walked by. Be prudent in the bush, especially near sun-facing rocks. If walking off cleared tracks wear leather boots, socks and jeans and ideally thick gaiters.

Figure 4.30. Red-bellied black snake: Pic: snake-catchers.com

If confronted by a snake, make no rapid movement. Especially don't wave things at it. Remain still, or slowly move away. Be particularly cautious during the breeding season of October/December.

The red-bellied black snake (Figure 4.30) is found in many parts of Australia. It is dangerously venomous but, as with most snakes, rarely bites unless provoked. Doctors say that almost all people bitten by snakes are male (often drunk), and usually as a result of pointlessly attempting to kill one.

If bitten, note the time (doctors need this). Do not wash off the venom: it will not seep into the wound and is needed for identification. Using a wide elastic bandage, bind firmly towards the extremity, and then towards the trunk. Immobilise the limb by strapping to anything suitable (e.g. a piece of timber) that is available. Avoid physical exertion and seek urgent medical attention.

Flooding

This is a serious risk in the upper part of Australia, especially throughout the wet season. When travelling in these parts carry ample food and water, and also a good reserve of cash. ATMs are few and far apart and not always accessible.

Figure 5.30. The author's OKA on Highway One Geraldton (WA) in 2000.

Most roads are officially closed when flooded, but there is a minor risk of being trapped along one if rain begins suddenly and heavily.

There are costly penalties for disobeying a closed road sign and you are likely to be charged a great deal for vehicle retrieval if you become bogged in this manner.

See also next chapter re cyclones.

Maps

Unless you drive only on the main highways, accurate and up-to-date maps are essential, but obtaining such maps can be a problem outside major cities. Some fuel stations carry a reasonable selection but they are not necessarily up to date.

In our experience, the best overall map of Australia is that sold by Australian Geographic. For local coverage the Hema, and also Westprint range is consistently reliable. It is also worth carrying one of the more comprehensive guidebooks. These usually include maps of most city and town centres.

If you have a laptop computer, a further possibility is the maps recently introduced. Printed maps are obtainable from major map shops (of which there are one or two in each major city).

On not becoming lost

Most of Australia's major tracks are well sign-posted, even across the Simpson desert, but station tracks can seriously mislead. If you intend to travel such routes carry a compass, and know how to use it. Keep a careful and ongoing note of distance travelled, checking for landmarks etc. that confirm your location.

Global Positioning System

Developed and provided by the US Department of Defense, the Global Positioning System (GPS) provides extremely accurate read-outs of position, height above sea level, speed and direction of travel.

Figure 6.30. A GPS unit is well worth having. The GPS function is also built in to many mobile phones. Pic: Garmin.

GPS receivers are well worth having if you intend to travel off main routes in isolated areas, where station and other tracks are often ill defined, and particularly so in the many large state forests.

Many people find GPS comforting in the outback but we recommend using it there primarily to confirm (or otherwise) your position based on conventional methods of navigation. Another and very real benefit of GPS is when driving in previously unknown towns and cities.

Finding north

If, in the southern hemisphere, and without a compass (but with a watch set to about the right time), lay it flat with 12 o'clock (non-daylight saving) pointing toward the sun. North (in the southern hemisphere) lies on that line halfway between 12 o'clock and the position of the hour hand - Figure 7.30.

Figure 7.30. How to find 'North. Pic: Wikihow.

Another way is to thread a needle through a cork and float it in water. Even if not magnetised the needle will point north and south. You can also often tell which is which even on a clouded day: north (in the southern hemisphere) is where the sky is brightest.

Personal locator beacons

The original Emergency Position Indicating Beacons (EPIRBs) were primarily for marine use but were used also on land. A Personal Locator Beacon (PLB) is preferable for RVs (and whilst hiking). The current system is digital and operates at 406 MHz. Each PLB transmits a unique identifying number and can be owner-registered on a central database. This enables rescue authorities to know whether the signal is from a boat, aircraft, bushwalker, etc. and to deploy resources appropriately.

The distress signal uses orbiting satellites, plus satellites in geostationary orbit over the Equator. This hugely speeds up the response time (from several hours - to minutes).

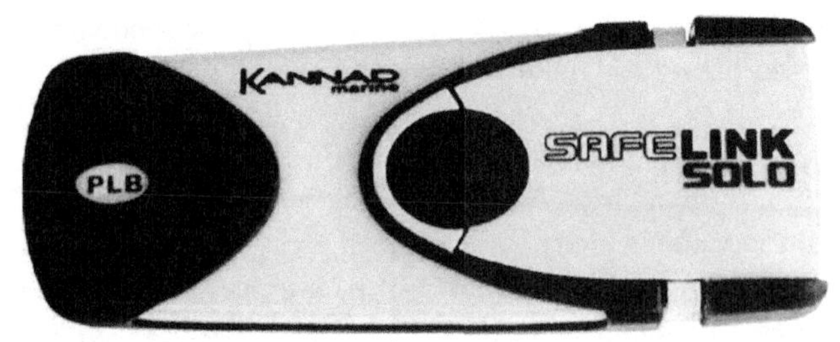

Figure 8.30. Personal Locator Beacon. Pic: RFD.

The beacon must have line of sight to the geostationary satellite (that will be low on the northern horizon for users in the southern states of Australia). Be aware that the signals will be blocked by rock outcrops and heavy foliaged canopies.

Whilst basic PLBs indicate position within a 5 km radius, the more up-market units have GPS capability than enables pin-pointing to within 120 metres - handy if you've accidentally stepped on that red-bellied black. The latest units weigh under 250 g and can fit into a coat pocket.

The beacons transmit a unique code that can be programmed to include who you are, and your vehicle details including registration number etc. Full details of how to do this are included by the manufacturer, but it also pays to ask the vendor.

A PLB should only be used as the last resort. In an emergency a satellite or mobile phone are far preferred for initial contact (you may be asked to activate the beacon to help pinpoint your location).

Remember to register your unit - beacons.amsa.gov.au/registration/index.asp#rego:

Information from: 1800 406 406 also www.amsa.gov.au/safety-navigation/search-and-rescue

Chapter 31

Cyclones

Most cyclones form north of Australia between November and the end of April. On average about ten tropical cyclones are forecast each year, of which six or seven eventuate.

They form and travel initially offshore, then cross the coast near Broome and Exmouth in WA, and Cape York and further south in Queensland. Some move west toward Darwin. Figure 1.31 shows the general pattern. Most cyclones lose strength and die out, but the survivors must be taken seriously. They often build up strength as they move, but progressively lose strength once away from the sea.

Only cyclones threatening life are reported by media, but there are typically three to four warnings each year for the north-west coastal area.

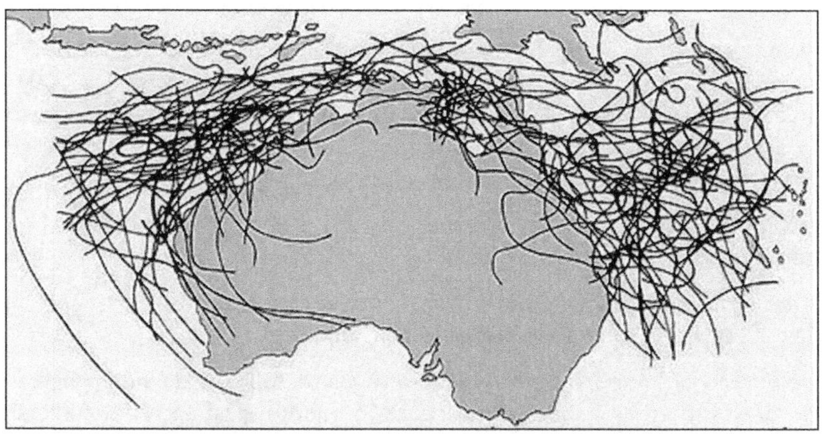

Figure 1.31. Paths of previous cyclones before dying out.
Map: Australian Bureau of Meteorology.

Essential preparation

Ideally, avoid likely affected areas during the cyclone season. If you decide to travel, however, prior preparation is strongly advised.

There is a rush for cash and supplies immediately after a warning. If a cyclone eventuates, electricity may be cut off resulting in ATMs, banks and stores closing. Keep fuel tanks topped up, and always have food and water to last seven days, and $300-$350 in cash.

Have a battery AM/FM radio, torches and spare batteries. Carry a map of the area that includes the ocean 500 km out. Although cyclones are far from 100% predictable this will assist you to plot their forecast movements, and to select exit routes.

Cyclone maps, plus *Surviving Cyclones* (published by Emergency Services Australia) are available at bom.gov.au/ SES offices, public libraries, councils and shire offices in areas likely to be affected.

Cyclone categories

Cat.	Strongest Gust	Typical Effects
1	<125 km/h	Damage to crops, trees and caravans.
2	125-169 km/h	Minor house damage, significant damage to signs, risk of power failure.
3	170-224 km/h	Roof & structural damage. Caravans destroyed. Power failure likely. (Severe Tropical Cyclone Roma.)
4	225-279 km/h	Significant roof loss & structural damage. Many caravans destroyed/ blown away. Dangerous airborne debris. (Severe Tropical Cyclone Tracy.)
5	>280 km/h	Extremely dangerous with widespread destruction. (Severe Tropical Cyclone Vance.)

Table 1.31. Official Australian cyclone categories. These describe the worst conditions.

Cyclone categories range from category 1 to 5. Each category describes the worst that may occur. Statistically, the effects you encounter are likely to be less. Cyclone activity reports are included in regular news broadcasts on local AM radio. Check this at least once a day or via the Bureau of Meteorology's automated telephone service (and bom.gov.au/).

Local police will be aware of cyclone activity. Local stores may display the latest cyclone map.

Other signs include frigate birds flying south, and snakes and lizards seeking shelter, but none substitutes for official information.

The Bureau of Meteorology's AM radio warnings follow a defined progression of risk: there are specific reports at 15 minutes past each hour, initially at three-hour intervals.

If cyclone risk increases, reports become hourly (at 15 minutes past each hour), prefaced by a strident siren warning signal. Reporting is then virtually continuous via local ABC AM - unless the only local station is FM - when it follows the same schedule.

Western Australia warning routine

The routine, starting with 'Cyclone Watch', covers any area (of Western Australia) within 100 km that might be affected if cyclone activity develops. If it does, the activity is formally defined as a cyclone, given an identifying name, and followed by a progressively graded series of warnings.

Blue Alert: a cyclone may affect the area but strong winds are not yet a threat. Evacuate now if you have not already.

Yellow Alert: a cyclone is moving close and appears inevitable. Strong winds are likely. Local authorities are likely to close all roads. If circumstances cause you to be there, seek assistance as soon as possible.

Red Alert: a cyclone and destructive winds are imminent. If wind forces/flooding permits, urgently seek closest assistance or the strongest and elevated shelter.

All Clear: signals the end of Red Alert. It is safe to go outside but proceed with caution because a Yellow or Blue Alert warning may still be current.

Other parts of Australia likely to be effected have generally similar sequences

Appropriate actions

If feasible, leave the cyclone area (which is likely to be the top of the Northern Territory, Queensland or north western Australia for 200 km or so inland) and head for the nearest settlement: most have public cyclone shelters. Ask police or the SES for advice. Do not panic or drive frantically, but don't delay as there is a high risk of prior torrential rain blocking roads and tracks.

If near the sea, storm surge is a real and serious risk. Storm surge is typically 2-3 metres, but can exceed 5 metres. If this coincides with peak tides (such as Broome's plus 10 metres), the surge is a mini-tsunami. The next biggest risk is flying debris.

Protecting your vehicle

Because they are light and high, campervans and fifth-wheelers are particularly vulnerable in cyclones. Motorhomes are vulnerable to a lesser extent. The accompanying drawings show how to secure a caravan, but similar principles apply to motorhomes.

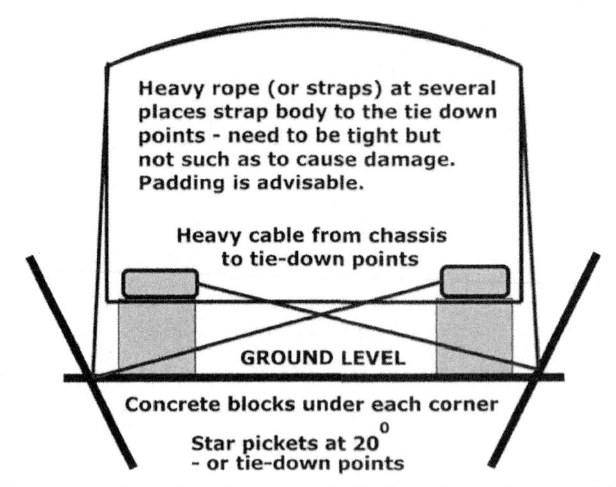

Figure 2.31. If possible, place concrete blocks under each corner. This limits the stress on the restraining ropes as is the RV cannot then roll on its springs.

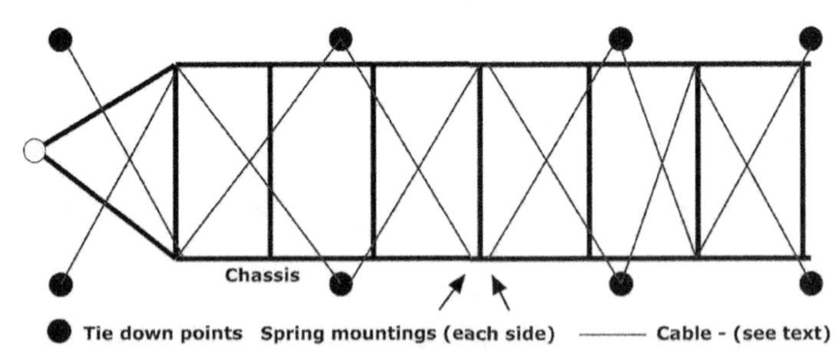

Figure 3.31. Secure the RV to tie-down points as shown here and Figures 3.31 and 4.31.

Place the vehicle with the nose toward the area's prevailing wind. If feasible, support the chassis on concrete or other blocks. If not, apply the hand-brake firmly and select reverse gear.

If 4WD, lock the front hubs (if relevant) and select low-range and first gear. Lower corner jacks (if they are fitted).

Secure the chassis to the supplied points or to star pickets driven deeply down at approximate 200 mm. (Figure 2.31).

If possible, use steel chain or cable with a breaking strength not less than 8 kN (1800 lbs). Secure loosely then tighten (using a length of wood as a twitch). Tighten rope to a twanging tension.

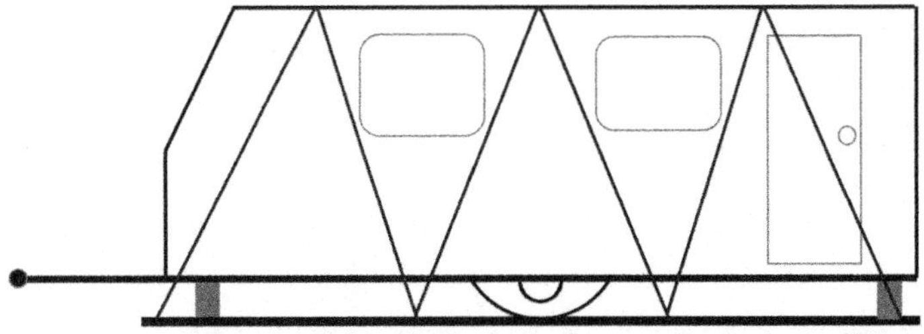

Figure 4.31. Tie down the body and roof - or they may be torn off the chassis.

Next secure the roof as shown in Figure 3.31 (totally essential with pop-tops) using the same strength rope. This rope needs to be secure but not to the point where it will damage the vehicle.

Run a further rope (not shown) horizontally at roof/wall level, inter-tying to each roof rope along the way to prevent any rope sliding out of position.

Place all loose items inside. Then leave. Never shelter inside the vehicle.

Chapter 32

Keeping well

People are living longer so it makes sense to benefit from that and enjoy it. How and why we age is not due to simply genetics: that accounts for only 25%. The remainder is up to us and our lifestyle. We need a diet low in fats and high in nutrients. Body weight, blood pressure and blood cholesterol need to be normal. We need regular exercise, ideally a lasting marriage (or partner) and, also ideally, to spend time with family and friends. We need social networks and active involvement in our community.

We need meaningful projects and things to be passionate about, plans, stimulating creative leisure, low stress, and optimistic and flexible thinking. We need to be open to change and to be able to live independently. We need to limit tobacco, alcohol and drug usage. The RV lifestyle partially assists, but more is really needed.

Learning, mental exercise and creativity

The human brain acts much like a muscle: the more we ask of it the better it performs. Learning literally regenerates new brain cells. Acquiring new skills, and understanding new ideas in different areas of thinking also opens opportunities to make new friends. The buzz we get from travelling, learning new things, different languages, creating our artistic mini-masterpieces, finishing that Sudoku or crossword puzzle keeps our brains sharp.

Nutrition

Healthy eating is essential: high calorie fat-laden food does not assist healthy aging. Good nutrition benefits the body and assists the brain. Eat moderately and cut back on meat, especially if processed. Eat fish, three serving of vegetables and two pieces of fruit a day. Limit alcohol to two standard drinks a day and have some alcohol-free days.

Exercise

'Sedentary individuals have a tendency to attribute the negative bodily changes they experience to 'aging' rather than their sedentary lifestyles' says Lakehead University's Brian O'Conner. This becomes a self-fulfilling prophecy. 'Regular exercise can prevent or dramatically reduce many of the physical and psychological declines that are commonly attributed to normal aging', says O'Conner. The simplest and most effective exercise is a combination of walking (not just strolling!), swimming and (particularly) active yoga.

Figure 1.32. Maarit Rivers, who contributed this chapter, taught herself to do this when she was 65. Pic: Collyn Rivers, rvbooks.com.au.

The former two are self-explanatory. Yoga really needs a good fully accredited teacher. 'Better a good book than a bad teacher', says the author (who over some 25 years personally taught and assisted to certify (via Yoga Alliance) over 80 of Australia's best.

Friends

We need many friends, and relationships with family and neighbours and the latter at least is usually enhanced by a travelling lifestyle. Pets too, boost well being. Any kind of social contact boosts our immune system and brain development.

Laugh more, live longer!

This, (a motto of the Campervan and Motorhome Club of Australia), has academic respectability! Successful aging really *is* aided by having a good laugh: it helps reduce stress.

Our capability to cope and move on helps us to live long and to be healthy and happy. This requires seeing problems from outside oneself - not to internalise them.

We need some excitement too. We need to take some risks. Novel intellectual challenges keep our brains sharp and boost immune systems. 'Only those who risk going too far can possibly find out how far they can go,' noted the poet T.S. Elliot.

Mind, body and spirit

Activities such as "music, imagery, meditation, yoga, tai chi, pet therapy, laughter and humour, a positive sex life and calm cheerful environment all assist", say Thompson, Sierpina and Sierpina, *What is Healthy Aging* (Generations 2002: 26; pp. 49-53).

Correct breathing increases your sense of well being. Learning to play the didgeridoo (from personal experience) combines the therapeutic qualities of learning something new, playing music and learning circular breathing. It is dramatically effective in reducing snoring (see also under Sleep Apnoea below).

Attitude is important. Those who say they are aging well are not necessarily the healthiest: but they have an attitude of optimism and good coping strategies.

Vices may help!

The odd vice speeds things up! - but one needs to select carefully. When feeling like staying in bed do so: have a good read and even a box of chocolates within reach. Our often loved (preferably dark) chocolate can reduce blood pressure, and the chances of having a stroke. But one needs to keep that chocolate in moderation or you'll be happy but seriously fat!

It is claimed that small amounts of red wine are better than beer - or none at all. But again, alcohol is taken to excess only at the expense of well-being and longevity. Pleasure and pleasurable feelings are good for us - releasing a chemical that is an antibiotic. Sex keeps us feeling good and staying young in mind and body. I suspect that for some, the caravan and motorhome way of life may go a fair way to assist.

Sleep apnoea

Reader feedback indicates a much higher usage of sleep apnoea machines amongst caravan and motorhome owners than in the general population. Why that is so is unclear, but for those who really do need such a machine away from mains power, rather than doing that via an inverter, consider one of the 12 volt machines now available. These use far less power.

If your medico agrees, people with mild forms might usefully explore the Buteyko method of breathing. Many have found it successful. Another fix (also for snoring) is playing the didgeridoo. Do not dismiss this - there are authoritative references. The currently best known is from the British Medical Journal and concludes: 'Regular didgeridoo playing is an effective treatment alternative well accepted by patients with moderate obstructive sleep apnea syndrome.' The peer-reviewed paper is well referenced and has been extensively cited by many learned journals. The full paper can be found at: bmj.com/content/332/7536/266

Becoming hurt/unwell

Consider completing a St John Ambulance (or equivalent) CPR (Cardio-Pulmonary Resuscitation) course. Better still complete the full two-day First Aid Certificate. Also invaluable is the Remote Area First Aid Field Guide, or Active First Aid (see also Table 1.32). The latter book, written by ambulance paramedics, is simple and concise.

Few commercial first aid kits are suitable for travellers: the latter book recommends making up the kit listed below. It is compact, yet adequate for routine and emergency situations.

(This chapter was written by the author's psychologist wife, Maarit Rivers, BA, MA.)

First aid kit

Adhesive strips	Splinter probe
Adhesive tape	Tissues
Alcohol swab	Triangular bandage
Antiseptic wipes	Tweezers
Combine dressing	Wound dressing (No 1)
Crepe bandage	
-(50 mm, 75 mm and 100 mm)	**Analgesics**
Gauze swabs	Antiseptic cream
Disposable gloves	Aspirin (also for inflammation)
First aid book	Burn cream (first use iced water)
Hand towels	Eye lotion
Non-adherent dressings	Eye pads (sterile)
Plastic bags	Lomotil or similar (for diarrhoea)
Safety pins	Panadol
Scissors	Water purification tablets

Table 1.32. Recommended contents of a First Aid kit.

Chapter 33

Where to stay

Many caravan parks began when caravans were small and lacked toilets, showers and need for electricity. Most were in coastal areas where families holidayed year after year. RV usage, however has long since outstripped facilities in peak periods. Without long prior booking, obtaining a site during such times in many areas is all but impossible. If you do, overcrowding may cause you to wish you hadn't (Figure 1.33).

Figure 1.33. Not necessarily typical - but hardly what many caravanners have in mind.

Despite caravan usage increasing, the number of caravan parks is declining: between 1997 and 2009 Queensland lost 63 (12% of all). Compounding the shortage, many caravan parks have since converted caravan sites to cabins that attract long-term tenants. Now, about 35% of a typical caravan park consists of such accommodation.

Caravan park ratings

Caravan park star-ratings do not relate to 'excellence' but mainly to facilities for amusing children, entertainment etc. Unless you need those facilities a two-star park is often a better and cheaper choice than a four-star park.

Apart from older or smaller RVs, many of today's are self-contained. They have an inbuilt toilet, usually a shower and clothes washing facilities - and many are independent of mains 230 volt power. They require a mere 40-50 square metres - but pay $30-$60 a night despite often cynical over-crowding.

Some caravan parks offer three nights for the price of two but do not always advise that, nor that they have cheaper sites if you do not need power. A few offer senior citizen discounts - or three nights for the price of two - but again you may need to ask. If you do stay in one, before booking in walk around and pick two or three sites that appeal. If you do not you will be allocated one that you may dislike.

Freedom camping

Despite the increasing inability to meet demand even in off-peak times, the caravan park lobby and many councils maintain that RVs should stay only in such parks. Not surprisingly, many ignore this. Surveys have shown that some 50%-60% of long term RV users free camp wherever possible - typically spending a day or two a week to do their laundry, shop, service the tow vehicle, etc.

Figure 2.33. The author's previously-owned 1974 VW Kombi (Wesfalia conversion) in a bush camp near Boulia (Qld). Pic: rvbooks.com.au.

Whilst there is little free-camping within about 25 km of Australia's east and south coast, there is still plenty elsewhere. Some sites are rarely-shared secrets, but many, particularly those along or close to major highways, are listed in books such as the (2019) *Camps 9*, and also on the Geowiki website. Contact details are in Chapter 38.

You may legally stay for prescribed periods in official Rest Areas in some states, particularly in Western Australia - where distances between caravan parks are huge. Most are listed in the camping guides mentioned above. Do not even think of staying in a truck parking area. It is illegal and noisy.

Where such areas are unavailable, it is usually necessary to go some distance off the road, often via tracks alongside a stream. Walk the proposed route first to ensure you can get out after heavy rain. If feasible, set up camp so you can drive straight out along a known firm exit. A four-wheel-drive tow vehicle provides comfort against bogging.

Many of Australia's innumerable national parks and state forests have variously basic facilities. Most can be accessed by an on-road RV. Pets are forbidden in all national parks and some state forests. Generators are usually banned or, where permitted, used only in an area otherwise reserved for (often noisy) tour groups.

Overnight fees are usually $7.50 to $10 per person, or less if there is no shower, but are increased from time-to-time. Some national parks also charge a substantial entry fee, but this is reduced for those with a Pensioner or Senior Citizen card.

Indigenous-run retreats/sites

Some outback national park and other campsites, in Cape York, the Northern Territory, and particularly on the Dampier Peninsular north of Broome, are run or owned by Indigenous communities. The many we have stayed in have been clean, well run, and with individual sites that are usually spacious and private.

You may also learn about the true Aboriginal culture along the way. Most of these sites are accessible by off-road caravans but some will need a four-wheel-drive tow vehicle.

Figure 3.33. Not many sites have views like this! The Goombargin Eco Retreat (north of Broome WA). Pic: goombaragin.com.au

Leave No Trace

Despite the number of caravan parks decreasing and an increasing number of self-contained rental cabins in those that remain, caravan parks lobby strongly and often effectively to prevent any form of free-camping. One of their strongest arguments is that many of those doing so leave behind rubbish (and worse) when they leave. This sadly is often only too true.

Chapter 34

What it costs etc.

The heavier the rig, and that which you load into it, the more it costs to run: not just for fuel, but for tyres, springs, shock absorbers, etc. Don't skimp on servicing, tyres or batteries. The cost of a breakdown usually far exceeds that of a missed or delayed service. If it happens in an isolated area it can cost thousands of dollars to have that rig transported to the nearest repair centre. In Australia's vast Kimberley area that centre can literally be 1000 plus kilometres away (a 2000 km return trip that may cost more than the RV is worth).

Fuel prices vary almost daily within cities, and are higher but more consistent in regional areas. Expect to pay 10%-25% more in many such areas, and double or more in remote outback communities. If space and weight permits, consider larger or extra tanks to avoid buying fuel where it is costly. This is particularly worthwhile for around-Australia travellers. Be wary of discount fuel.

LP gas is less readily available and costs twice that of city prices in most of the north and north-west.

Fuel costs are reduced by driving less and walking (and seeing) more. For large petrol-engined vehicles, fuel consumption in town increases far more than for diesels. On sand it may double. Keeping to 80 km/h cuts fuel usage by 25%. Where choice is available, consider caravan parks with the lower star ratings. As noted in Chapter 33, such ratings relate almost entirely to children's' entertainment and other facilities that most RV owners do not want or need.

Food

Food and caravan park fees tend to be the major costs. The former can be reduced by cutting down on red meat and buying fish only near the coast, a major river or catching it yourself. You need a fishing licence in most states - see Chapter 35 re the often huge penalties if you are caught fishing without one.

Use rice, pasta, couscous, potatoes and bread as your main source of carbohydrate. Where/when feasible, buy only fruit and vegetables that are in season. Roadside stalls may have them at bargain prices, as may markets just before they close. Stock up on specials for basic goods. Ice creams, biscuits and soft drinks blow budgets apart (and increase waistlines). A few country towns have really good restaurants, but basic meals can usually be obtained cheaply at local clubs.

Alcohol

Much cask wine is dreadful but some good stuff is identical in both bottles and casks. Try 'clean-skin' wines', they are usually surplus quality products that their makers do not wish to be identified, as it prejudices full-price sales. Many towns in the Northern Territory and Western Australia impose quantity restrictions as well as time of day restrictions on buying alcohol. The sale of wine casks is forbidden in northern WA.

General

Shop around for insurance. Various companies quote different prices for apparently similar cover. Premium services offered by motoring organisations provide extended cover for a higher price.

People on the road tend to remain healthy and incur lower medical bills: $50 a week typically covers minimal insurance and minor medications but some may pay more.

Many towns (particularly in Western Australia) have op-shops where paperbacks sell for 10-20 cents. Some public libraries have a free paperback exchange. There are also commercial book exchanges - but costly for avid readers. Video/DVD hire is cheap in country towns but, as with those from caravan parks, the choice can be deadly.

Clothing chains unload excess stock in remote country areas at ultra-low prices. High quality linen dresses can often be bought for $10 to $15 in the Boulevard Centre's shopping mall in Broome. Bargains are found in op-shops: new King Gee overalls for $15, a new hand-knitted sweater for $4.

Most outback stores, hotels and fuel suppliers accept credit cards, but a few accept only cash. Cheques are rarely accepted. Carry cash in smallish denomination notes, or a bit larger if you have a thirsty vehicle. Most towns have ATMs, and supermarkets will provide cash drawn against an EFTPOS card.

Concession cards

Holders of Pensioner, Senior Citizen, and Commonwealth Senior's cards receive concessions on council and water rates, vehicle registration, driving licences, and telephone fees. Details can be obtained from any Centrelink office (or 1800 671 233).

Selling up

Some intending long-term travellers sell up - and then miss having a permanent base. Meanwhile house prices escalate. If you sell, you may not be able to afford to buy again. Consider having a real estate agent let and manage your home before committing to a very different and yet untested way of life.

Table 1.34 originated in the *Campervan & Motorhome Book* in 2001. It has been updated many times since. It is mainly applicable to those who travel the current Australian average of about 13,500 km a year. There are indications that a growing number of people adopting a long-term travelling lifestyle are covering shorter distances and driving less often, but mainly saves cost on fuel.

Apart from initial vehicle costs, the ongoing expenses of caravan and campervan/motorhome owners are much the same. The yearly cost for many, however, is less than those who live at home as many such spend a lot on air fares and accommodation for their annual holidays.

A still ongoing trend as RVs become increasingly self-contained, is an ongoing reduction of caravan park usage. Site fees have risen so much, however, that despite such less usage, the amount spent is now higher.

Vehicle depreciation is not included but that for caravans is considerably less than for campervans and motorhomes. Many claim it to be virtually zero once it is over five or so years old. Loss of interest in capital tied up in the rig is not included, but accountants would probably insist it is. Licence renewal, presents, etc. are not included.

Telephone, entertainment and dining out costs vary considerably.

The average costs for two people driving a motorhome or towing a caravan via a 4WD and travelling a typical 13,500 km in one year is likely to be (until 2021) from $32,500 to $50,000, but be aware that fuel in outback areas can be at least 50% higher than in towns.

The table is less applicable for those travelling extensively off-road as fuel costs, tyre wear, vehicle maintenance and repairs inevitably cost more.

Communications

Mail can be sent for subsequent collection to any post office anywhere in the world. The official way of addressing such mail is as follows:

Your name,

POST RESTANTE, 'Wherever'

Post Code

Professional services provide reliable and speedy mail forwarding but, as they will not pre-sort, you pay heavily for junk mail. This issue is reducing as more and more such mail becomes electronic.

One often overlooked issue is that, when forwarding is no longer needed, unless payment continues subsequent mail is returned to the sender. This is not unreasonable, but is costly. The fastest affordable bulk mail is via Australia Post's prepaid 3 kg and 5 kg Yellow Express satchels.

Delivery is overnight in some areas, and usually within three working days even if sender and receiver are outside the Express Delivery area. Satchels are tracked as they travel through the system.

Item	$Weekly	$Yearly
Rig costs		
Registration		600
Insurance		1100
Service & repairs		1000
Fuel/oil		4500
Tyres		550
Sub-total		7250
Living costs		
Food	190	9880
Eating out	75	3900
Site fees	120	6240
Entertainment etc.	70	3640
Wine/beer	40	2080
Telephone	10	520
Shoes/clothing	15	780
Laundry	30	1560
Newspapers/magazines	15	780
Medical (for two)	50	2600
Gas	20	1040
Postage	10	520
Sub-total		33,840
Total yearly spending		41,190

Table 1.34. Typical yearly costs (late 2019).

Chapter 35

Legal issues

The wide range of RV vehicles available has an equally wide range of legislation relating to it, and to the license class needed. Those for camper-trailers, caravans and fifth-wheel caravans differ in some respects from those for campervans, motorhomes and coaches.

Tare Mass

The Tare Mass is the unladen weight of an RV as sold. For camper-trailers and caravans Tare Mass includes water tanks but *not* the (1 kg/litre) weight of water. It includes an 8.5 litre LP gas bottle, but not its contents.

For campervans, motorhomes and coaches, Tare Mass includes the weight of filled water tanks plus ten litres of fuel. The Tare Mass is legally required to be shown on a Compliance Plate attached to the vehicle. (In this section's context, 'mass' can be regarded as 'weight'.)

Most RV buyers assume the maker installs 'optional extras' such as air-conditioners, extra batteries, etc. specified in the purchase order. Some do, but many make 'standard basic products, and have the dealer supply and install those extras. None of the extras (and never the 100-200 kg of water and the 8.5 kg of gas) is likely to be included in the Tare Mass shown on the compliance plate. The 'unladen' weight may thus be far higher than expected and can seriously degrade the RV's intended usage.

Buyers should ensure the purchasing contract includes a specified weight allowance. The RV's weight should be checked and recorded on a certified weighbridge in your presence prior to accepting delivery.

Aggregate Trailer Mass

Also known as the Gross Loaded Mass, this is a legally obligatory Rating. It is the total allowable mass of a laden trailer when carrying the maker's recommended maximum load. It includes that imposed on the tow vehicle when both trailer and tow vehicle are on a level surface.

Unlike campervans and motorhomes, camper-trailers, conventional and fifth-wheel caravans have no legally obligatory minimum allowance for personal effects. There is only a de-facto industry recommendation of 250 kg for single-axled units and up to 450 kg for dual-axled units (custom-made and off-road trailers may have more). That weight of water in full tanks reduces that by 100-200 kg.

Gross Trailer Mass

The Gross Trailer Mass is that borne by a trailer's axle/s when carrying its maximum permitted load uniformly distributed across its load bearing area. It is the ATM minus the nose weight imposed on the tow bar. (The GTM is usually specified for commercial load carrying trailers.)

Gross Combination Mass

The Gross Combination Mass (GCM) determines the maximum on-road weight of any vehicle towing a trailer, plus the laden weight of that trailer. It is thus the GVM plus the ATM.

Gross Vehicle Mass (GVM)

The GVM is a Rating that applies to campervans, motorhomes and converted coaches and tow vehicles (but not trailers). It is the actual Tare weight plus everything you carry including the weight of the driver, passengers, fuel and water, i.e. it is the maximum that RV may legally weigh on-road when fully laden.

The Tare Mass (for motorhomes) includes filled water tanks but only 10 litres of fuel. The GVM includes the weight of the driver and passengers, plus any obligatory personal allowance of 60 kg for each of the first two sleeping berths and 20 kg for each thereafter. The personal allowance is intended to cover bedding, cooking utensils, food and luggage. Many owners claim it to be grossly inadequate - particularly in the small/medium size campervan area. See Chapter 36 for typical weights.

The GVM is determined by whichever is the lowest of the maximum permissible tyre loading, axle loading and the vehicle manufacturer's specified maximum weight limit. The load you can legally carry is thus the difference between the 'true' Tare Mass, and the rated GVM.

Manufacturers must ensure the obligatory personal allowance is provided. If you had expected more, yet failed to specify that you needed that (and by how much) when ordering, both manufacturer and dealer are likely to insist it is not their problem.

It may be possible to have the GVM rating increased but may be costly if it requires the whole suspension, axle/s brakes, wheels and tyres to be replaced. A full engineering report is required.

Legal towing weights

Australia's national rules specify that for vehicles with a GVM not exceeding 4.5 tonnes, the maximum permissible trailer weight (including load) is the lesser of the capacity of the towing apparatus, or the relevant maximum trailer mass specified by the tow vehicle manufacturer.

The above partially overrides the 1998 weight legislation that (where such information could not be obtained) the maximum permitted trailer mass is one and a half times the unladen mass of the towing vehicle, if the trailer is fitted with brakes, or the unladen mass of the motor vehicle if the trailer is not fitted with brakes. It is expected that new legislation may eventually bring local practice into line with the more stringent requirements of Europe.

Weighing-in stations

Load-carrying vehicles above certain weights must pull into weighing stations for checking that they are within their rated weight. Requirements vary from state to state. Notices advising the vehicles affected are displayed at the entry points.

Driving licence requirements

Holders of a C class driving licence may drive any vehicle with a Gross Vehicle Mass of less than 4.5 tonnes (and with seating for up to 12 adults). They may also tow a conventional or fifth-wheel caravan with such weight not exceeding the Gross Combination Mass. The previous exception relating to fifth- wheel caravans registered in the ACT was brought into line with that above in 2014.

An LR driving licence is required for vehicles of a Gross Vehicle Mass (GVM) exceeding 4.5 tonnes and less than 8 tonnes. The requirement relates to potential carrying capacity. If the vehicle's GVM is 5.5 tonnes, but even if it is never loaded beyond 4.49 tonnes, you still need an LR driving licence. Vehicles above 8 tonnes require an MR or HR licence driving licence.

Licence and registration renewal

Renewing a driving licence from outside one's 'home' state can present unexpected difficulties. The main problem is that you must provide a street address. Arguing that you do not have one is useless. There is no choice but to use a friend's or relative's address. No address - no licence. Vehicle registration is likewise. Here again you must use someone's street address (this address need not have a garage or even a place to park a car!). The reason for its need is that the official bodies concerned require a physical address that enables them to contact you by mail if deemed needed.

Some authorities will renew a driving licence and/or registration before their expiry date/s. If this works for your planned trip this is the best way to go. If this is not feasible, establish the current procedures by contacting the relevant authority in your home state. Do this before you leave as most list 013 telephone numbers that can only be dialled from within your home state. Directory enquiries cannot provide their direct numbers. The information can, however, usually be found on the Internet.

Personal Property Securities Register

The Personal Property Securities Register (PPSR) advises whether used vehicles (including RVs) have any money owing etc. Dealers are now legally obliged to provide a PPSR certificate (valid on the day of delivery), showing that no encumbrances exist. If buying privately it is strongly advised to obtain a PPSR certificate confirming clear title before you part with your money. See ppsr.gov.au

Import issues

Vehicle Standards Bulletin (VSB) 1 'Building Small Trailers' permits manufacturers and importers of small trailers to self-certify compliance of light trailers, enabling them to by-pass the certification requirements obligatory for other categories of new vehicle.

In 2007, the federal government officially warned that some importers falsely understated (mainly fifth wheeler) trailer weights to enable them to be illegally registered in a lighter category. Safety implications included inadequate braking and overloaded suspension and tyres. Checking compliance of these and other imports is now increasingly enforced, particularly if re-registering in a different state. Caution is particularly needed with RVs imported prior to mid-2010 (and that changes may apply as from 2020). The reference is: infrastructure.gov.au/roads/vehicle_regulation/bulletin/trailer-importers.aspx.

Private imports

Legislation enabling private imports may involve 'facilitated importing' that results in the legal transaction being between buyer and overseas seller. Whilst 100% compliance is not required if the unit is used in Australia only by the original buyer, it cannot legally be resold unless made 100% compliant.

Unless a legal contract binds the facilitator, liability lies with the original buyer. It can be sold to a dealer, but that dealer automatically incurs the liability. For some vehicles (such as those that are overwide) total compliance is not possible.

Camping defined

Authorities define camping in various ways but it is the local council's health department that usually prohibit the overnight occupied parking of an RV in defined areas - whether the vehicle is self-contained or otherwise. In practice you are unlikely to be disturbed if you stay overnight in a quiet back-street not overlooked by nearby residents. Do not raise a pop-top roof, nor let anything drain onto the ground.

Don't stay overnight near an established caravan park, and absolutely do not cheat caravan park owners by using their toilet and other facilities without paying. Some people do.

Drinking alcohol whilst camped

Under Section 25 of the Police Offences Act it is illegal to drink alcohol in a public street or public place (prescribed by regulation for the purposes of Section 25). Subsection 25 (5) extends this to make it an offense to drink alcohol in a stationary motor vehicle that is on a public street, or (prescribed) public place.

The intent of this part of the Act is (to quote from Hansard, 10 May 1995) 'to cover those people who move from hotels to their cars, with liquor, and sit drinking their liquor. Often after consuming the liquor, they return to the hotel, purchase more, and on the way to and fro, damage property and otherwise conduct themselves in an antisocial manner.'

It is technically an offense to consume liquor an RV even if settled down for the night in a public street, or (prescribed) public place, but most police are aware of Act's intent. It is unlikely that police would take any action in the above circumstances, particularly if the RV's engine starting key was not in the ignition. In the improbable event that they did, any consequent charge would likely to be thrown out of court.

Fishing licences

Fishing licensing and fishing restrictions vary from state to state and from time-to-time. Take these very seriously. Illegal fishing can (e.g. in the Northern Territory) result in confiscation of all associated equipment. This is not just confined to your rods and lines - it may include your rig and its contents.

Voting

Even though you may be away from home for some length of time, or spend life more or less permanently on the road, Australian law still requires you to vote. The simplest way is to preregister as an 'itinerant voter'. This status enables you to cast your vote from any polling booth, anywhere.

Any Electoral Office can assist you to do this. Their locations can be found by Googling - and are listed under that heading in Telstra White Pages telephone books.

Chapter 36

Approximate weights (kg)

General		Water	
Awning	20	*General + drinking 120*	
- surround	20	Supplies-various	
- poles (6–12)	15	Firelighters	
- ropes/pegs (8–16)	5	Matches	
Binoculars	2	Paper towel	
Blankets/doona	10	Spare plastic bags	
- pillows (2)	2	**Sub-total**	**2**
Bush saw	1.5		
Camera	1.0	**Cleaning**	
Camera lenses memory cards		Dishcloth/s	
Chairs (4)	6	Garbage bags	
Clothes line (20 m)	0.5	Hand broom/pan	
Clothes pegs (50)		Scrubbers	
Computer (laptop)	2.5	Spare bucket	
- printer	1.5	Sponges	
- paper	1.0	Steel wool	
Doormat	1.0	Washing up detergent	
DVD player	2	Wipes	
Fire extinguishers (2)	3	**Sub-total approx.**	**6.0**
Fly nets (2)			
Funnel/s (2)	0.5	**Clothes**	
Machete	3	Hat (large brim)	
Mains supply cord	3	Jacket (warm)	
Mattress foam (double)	8	Jeans	
Mobile phone		Raincoats (plastic)	
Satphone	0.3	Shirts	
TV (34 cm)	5	Shoes	
Shovel (small)	1.5	- general (1 pair)	
Spade	1.5	- hiking (1 pair)	
Tables (2)	5	Shorts (rough)	

Tarpaulin (plastic)	5	- (decent)	
Torch	0.3	Slacks (2)	
- globes/batteries)	0.5	Socks	
Sub-total	**127**	Sweaters	
		Swimmers	
(Optional) tent		Tee shirts (7)	
Tent & fly sheet 10 guys & pegs	3	Thongs/sandals	
- mallet	1	Underwear	
- mattress air	2	- thermal	
- light (electric)	0.5	**Sub-total (for two)**	**14**
Sub-total	**16.5**		
		Personal	
Kitchen - main		Back packs	
Barbecue plate	5	Contact lenses	
Camp-oven	3	- cleaners	
Gas 9 kg (2)	40	Massage oil	
Outside cook top	3	Mirror	
Sub-total	**51**	Nail file/s	
		Razor	
Kitchen - general		Safety pins	
Bottle opener		Sewing gear	
Bowls (4)	1	Soap	
Can opener	0.2	Sunburn cream	
Chopping boards (2)	1	Sunglasses	
Egg cups		Tampons	
Kettle	0.2	Toilet rolls	
Knives	1.0	Towels	
- bread	0.5	**Sub-total (for two)**	**5.0**
- paring	0.8		
Mugs (4)		**Medical**	
Plates (4 large) 1.5	0.5	First aid kit and basic medicines - (page 85) [2]	
- small (4)	1.0	Recreation	
Saucepans (3)	2.0	Books (paperbacks)	
Spatulas		Compact Disks (CDs)	
Spoons (wooden)		Magazines	

Thermos	1.0	Painting/drawing gear	
Sub-total	**11**	Paper	
		Pencils	
Food (basics)		Swim gear	
Biscuits		Water bottles	
Bouillon cubes		**Sub-total approx.**	**15**
Bread			
Butter		**Tools**	
Chocolate		**Sub-total**	**15-50**
Coffee			
Cooking oil		**Vital things**	
Cold drink/s		ATM Cards	
Flour		Cash	
Fruit		Cheque book	
- fruit tinned		Driving licences	
Jam/marmalade		Insurance papers	
Milk		Tax returns data	
- milk UHT		**Sub-total**	**1.0**
Mustard			
Noodles		**Reference**	
Oatmeal		Bird book	
Rice		Calendar	
Salt/Pepper		Guide books	
Soups		Instruction manuals	
Spaghetti		- vehicle	
Spices		- general	
Tea bags		Maps	
Vegemite		Medical guide	
Vegetables		**This book!**	
Water		**Sub-total**	**9.5**
Wine			
Sub-total approx.	**20**	**Total weight**	**400-500 kg**

Table 1.36. Note that the weight of water (plus 10 litres of fuel) is included in the Tare Mass of campervans, motorhomes and coaches, but not in the Tare Mass of caravans. Kitchen, food, clothing and personal sub-totals are approximate. It is assumed that saucepans etc. are stainless steel. Use melamine, dishes and plates, etc. as (stainless steel plates cool food rapidly). The weight of tools will vary considerably, but it is advisable to carry some: if you do not know how to them, someone who can is likely to then assist.

It is unlikely that all listed will be needed but if spending time on the road, overall weight will be much as above. As can be seen the personal allowance provided for many production RVs is grossly inadequate.

Chapter 37

About us and our vehicles

Figure 1.37. The QLR demonstrating its climbing prowess.
Pic: rvbooks.com.au.

My experience in this field began in the late 1950s whilst working for the Vauxhall/Bedford Research Laboratory. Much of my work and interest involved suspension behaviour on corrugated roads.

Africa was a major Bedford market, but not enough was known about its road surfaces to improve our test methods, so (then in my late 20s) I offered to take a Bedford across Africa a couple of times to check. Spotting an ulterior motive, the chief engineer felt that Bedford could not offer 100% support but assisted to 'liberate' an unused QLR 4WD truck and also persuaded Mobil to fuel the vehicle throughout.

The QL range was initially of military vehicles for WW2. Their 7 tonne (laden weight) was grossly underpowered by a 3.5 litre petrol engine of 54 horse power (less than that a 2019 Suzuki Jimny). This limited them to a maximum speed, on flat roads) to under 50 km/h, however their ultra-low gearing (bottom was 101:1) enabled them to virtually climb the side of a house - at about 1 km/h.

The QLR version was rare - it was built primarily as a Royal Air Force mobile airfield control centre and had a 12 volt 400 amp dynamo driven from the main engine via a power take-from the gearbox.

Dunlop donated six 1100 by 20 tyres - surprising us by recommending standard tyres (used also by London buses!). They were totally worn out after the plus 70,000 km trip but performed superbly.

A year or so was spent making its coach-built body vaguely livable, adding specialised instrumentation, plus fuel tankage sufficient for a minimum 3500 km range (needed for the Saharan crossings).

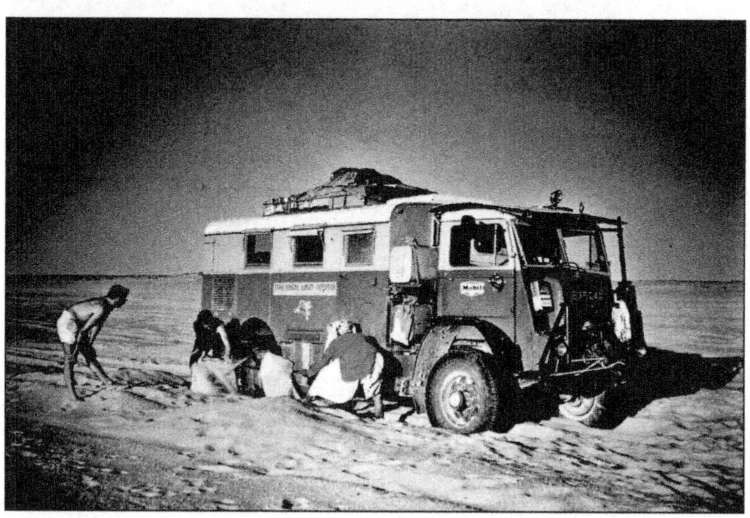

Figure 2.37. The QLR temporarily bogged mid-Sahara. Collyn 'supervises' its recovery. The man on the right is Mahatmadu, a Toureg who'd hitched a 1100 km lift. Pic: Tony Fleming.

The southern crossing took about three weeks and, apart from having to drive through what was then a war zone, and staying within French military compounds at night, was relatively easy. The track back then was rough, but required four-wheel-drive only occasionally until the (then) French Foreign Legion oasis/town of Tamanrasset.

From there-on there is about 1000 km of often soft sand. There is no 'track' as such: the general direction is shown by a wooden post every 10 kilometres. We became bogged a few times, but never that seriously.

Our major issue (apart from avoiding riots etc.) was traversing Africa also east-west as this had to be done during the rainy season.

This necessitated 4WD plus steel chains on all wheels for much of the east-west crossing.

We came close to rolling the QLR a few times, but managed to retrieve it each time using our trusty Tirfor manual winch.

Figure 3.37. Our 20 year old QLR pulled this broken down Jeep through 750 km of soft sand on the northern return part of the trip. Pic: Tony Fleming.

The return Saharan crossing was far harder.

We had foolishly offered to tow a then 12 months-old Forward Control Jeep virtually the whole 3000 plus kilometres from Kano (Nigeria) to almost Algiers.

The Jeep had broken almost one of everything, including the drive to the rear wheels (and its front wheel drive was of next to no value in soft sand).

We did however, by now have a better feeling for avoiding sand that was softer than it seemed, but to safeguard the QLR's transmission, we frequently stopped and with the QLR as an anchor used the Jeep's still-working power winch to pull it through the softer patches. (Figure 3.37).

During our return Africa politically exploded behind us. The Saharan route was closed on the day we left and has not been opened since. The QLR was the very last vehicle to this day to make the trip through Africa's centre. A superb video of this epic trip can be viewed at: https://rvbooks.com.au/page/about/

Figure 4.37. Fortunately we had foreseen the need for adequate fuel filters! Our Kombi taking on fuel in the outback town (Boulia in 1997). Pic: Caravan & Motorhome Books.

Following that trip I migrated to Australia. In 1993 my wife (Maarit) and I bought a 1974 Wesfalia VW Kombi (Figure 4.37) and rebuilt the suspension to make it more suitably off-road. It proved ultra-reliable and in that form travelled across Australia via dirt tracks for over 35,000 km.

Whilst far more rugged than it may appear, a Kombi's low slung rear engine limits its wading depth. Furthermore its end weight makes it very hard to keep in a straight line on soft muddy roads.

The Kombi was also our first experience of using solar. It had a horizontally mounted 80 watt solar module that, in conjunction with a 100 amp hour lead acid battery, provided ample energy for our halogen lighting.

Figure 5.37. The caravan at Lake Inarijarvi (far north Lapland). Pic: (at midnight!). Pic: rvbooks.com.

We later borrowed Maarit's brother's then 40 year-old 15ft caravan. We towed it from Helsinki to Norde Cape (Norway) the furthest north one can travel by road. It is about 750 km north of the Arctic Circle, and 2000 km from the South Pole.

OKA

In 1994 we bought a nine months old OKA ex mining vehicle. It had a coach-built 14-passenger body that we converted into a fully off-road motorhome.

After a trip to Darwin, via the Simpson desert, we returned to Sydney via Broome. Whilst there, we bought 10 acres of Indian Ocean water frontage north of Cable Beach. We drove back to Sydney via Alice Springs, sold up and returned to Broome (via Cape York) taking some 18 months to do so.

The OKA provided our home whilst we physically built an all concrete, steel and cyclone-proof glass home. It was totally solar-powered via a 3.4 kW stand-alone system that we also designed and built ourselves.

From 2000 to 2006 we took the OKA across Australia and back (about 13,000 km) some 12 times using dirt tracks for over 70% of the way, and also to take visitors innumerable times along the Gibb River and Cape Leveque Roads.

As with most OKAs we had early teething problems (mainly suspension-wise) but once sorted out it was totally reliable and superb off-road.

We sold the OKA in 2007, replacing it by one of the very last made 4.2 litre turbo-diesel Nissan Patrols, and a TVan (that I had fitted with disk brakes).

As with the OKA, the Nissan Patrol was reasonably quiet, comfortable and thoroughly reliable. We fitted locking front hubs, a 7000 kg electric winch and a snorkel.

Figure 6.37. The OKA crossing the Simpson desert's major sand dune - 'Big Red'. Pic: Maarit Rivers.

It also had the very first Redarc BMS dc-dc alternator charger (on long term trial), plus solar modules on its roof that powered a 60 litre Engel fridge in the rear of the Nissan Patrol.

The TVan is tiny, virtually a bedroom on wheels plus a slide-out kitchen. travel ultra-light). Towed with ease by the Nissan Patrol, the rig was driven from Broome (via Australia's centre) over mostly dirt tracks to the east coast and back three times.

We'd give it a thorough service before leaving, do an oil change half-way and then it drive back to Broome.

Neither the Patrol nor the TVan required any repairs, nor had tyre problems.

We finally moved back to Sydney, in late 2010, to be closer to our family, and Maarit's desire to do her (now completed) MA, plus our publishing business becoming too big to run from Broome.

Home is of course all-solar - including Tesla battery back-up.

Figure 7.37. Our previously owned 4.2 litre Patrol on the track to the Mitchell Plateau (in the Kimberley). Pic: rvbooks.com.au.

Chapter 38

Contacts & links

RV organisations

Australian Camper-trailers Group: group for camper-trailer owners (membership is free). Whilst free of commercial pressure it also welcomes related manufacturers and suppliers to join and share their knowledge and experience as long as it is unbiased and free of promotion: campertrailers.org/

Big Rigs Clan: a special interest group of the CMCA (above) catering for those with buses or coaches over 10.5 metres self-converted to motorhomes: bigrigsclan.com/

Campervan and Motorhome Club of Australia: with about 70,000 members, the largest Australian club for RV owners. Member benefits include publications, insurance, rallies etc: cmca.net.au/

Highway Wanderers: CMCA special interest group for those travelling full-time: highwaywanderers.com/

Solos: yet another CMCA initiative for those travelling alone in RVs. Primarily female membership but males welcomed. Many members form small sub-groups and travel together: solosnetwork.org.au/

Caravan Club of Australia: founded in 2006 currently has 4000 members: australiancaravanclub.com.au

Caravan Council of Australia: a non-profit RV-owner oriented organisation run primarily by engineers provides unbiased technical advice on all RV issues: caravancouncil.com.au

Caravan Industry Association of Australia: Caravan Industry Association of Australia is the peak National Body for the Australian caravan and camping industry representing over 4000 industry businesses. It also represents the Australian caravan park industry: caravanindustry.com.au/

RV publications and websites

Australian Caravan Park Reports: website with unbiased, up-to-date reports on the quality of specific caravan parks Australia-wide. Each caravan park is rated by caravanners: acpr.myparklist.com/

Camps 10: Long established and routinely updated guide to caravan parks and free-camping sites in Australia: campsaustraliawide.com

Caravan & Motorhome Electrics: by Collyn Rivers. All you ever need to know about this subject. Most of Australia's auto electricians buy it too: rvbooks.com.au.

Solar That Really Works!: by Collyn Rivers. A complete guide to all aspects of installing and using solar for boats, cabins and RVs.

Why Caravans Roll Over - and how to prevent it: by Collyn Rivers. Here, for the very first time is a plain-English explanation of this still increasing phenomenon. It also includes a fully referenced technical explanation.

RV Books: our associated website has a large number of constantly updated technical and general articles on every aspect of RVs and their use: rvbooks.com.au/

iMotorhome: free eMagazine about motorhomes and campervans in Australia and New Zealand. Free download on the first and third Saturdays of each month. imotorhome.com.au/

No Boundaries: a website for RV travellers in Australia. noboundaries.com.au/

On The Road Magazine: Australian monthly magazine for caravanners, camping enthusiasts, motorhomers, 4WD, outdoors and eco-tourism. ontheroadmagazine.com.au/

RV Pages: supported by CMCA, RV Pages is a comprehensive guide to vehicles, accessories, services and destinations for RV enthusiasts across Australia. rvpages.com.au/

Useful RV service

Global Gypsies: employment/volunteering worldwide for RV travellers : globalgypsies.com.au/nomads/

Table of Contents